Manufacturing Competitiveness in Asia

There are competing theories to explain the reasons behind the international competitiveness of manufacturing in Asia. Analysing these different theories will bring important lessons, not just for Asia, but for developing economies the world over.

This lucid book studies industries and firms in East Asia and examines the major determinants of their economic performance. With contributions from leading thinkers, including Ha-Joon Chang and Rajah Rasiah, the book covers such themes as:

- industrial policy and East Asia
- Taiwan's information technology industry
- the role of the government in technological capability building.

Manufacturing Competitiveness in Asia touches on many important themes and issues and as such will be of great interest to students, academics and policy-makers involved in industrial economics, international trade and Asian studies.

Jomo K.S. is Professor in the Faculty of Economics and Administration, University of Malaya, Kuala Lumpur, Malaysia. Another of his books, *Southeast Asian Paper Tigers?* is also published by RoutledgeCurzon, in 2003.

RoutledgeCurzon Studies in the Growth Economies of Asia

1. **The Changing Capital Markets of East Asia**
   Edited by Ky Cao
2. **Financial Reform in China**
   Edited by On Kit Tam
3. **Women and Industrialization in Asia**
   Edited by Susan Horton
4. **Japan's Trade Policy**
   Action or reaction?
   Yumiko Mikanagi
5. **The Japanese Election System**
   Three analytical perspectives
   Junichiro Wada
6. **The Economics of the Latecomers**
   Catching-up, technology transfer and institutions in Germany, Japan and South Korea
   Jang-Sup Shin
7. **Industrialization in Malaysia**
   Import substitution and infant industry performance
   Rokiah Alavi
8. **Economic Development in Twentieth Century East Asia**
   The international context
   Edited by Aiko Ikeo
9. **The Politics of Economic Development in Indonesia**
   Contending perspectives
   Edited by Ian Chalmers and Vedi Hadiz
10. **Studies in the Economic History of the Pacific Rim**
    Edited by Sally M. Miller, A.J.H. Latham and Dennis O. Flynn
11. **Workers and the State in New Order Indonesia**
    Vedi R. Hadiz
12. **The Japanese Foreign Exchange Market**
    Beate Reszat
13. **Exchange Rate Policies in Emerging Asian Countries**
    Edited by Stefan Collignon, Jean Pisani-Ferry and Yung Chul Park
14. **Chinese Firms and Technology in the Reform Era**
    Yizheng Shi
15. **Japanese Views on Economic Development**
    Diverse paths to the market
    Kenichi Ohno and Izumi Ohno
16. **Technological Capabilities and Export Success in Asia**
    Edited by Dieter Ernst, Tom Ganiatsos and Lynn Mytelka
17. **Trade and Investment in China**
    The European experience
    Edited by Roger Strange, Jim Slater and Limin Wang
18. **Technology and Innovation in Japan**
    Policy and management for the 21st century
    Edited by Martin Hemmert and Christian Oberländer
19. **Trade Policy Issues in Asian Development**
    Prema-chandra Athukorala
20. **Economic Integration in the Asia Pacific Region**
    Ippei Yamazawa
21. **Japan's War Economy**
    Edited by Erich Pauer
22. **Industrial Technology Development in Malaysia**
    Industry and firm studies
    Edited by Jomo K.S., Greg Felker and Rajah Rasiah
23. **Technology, Competitiveness and the State**
    Malaysia's industrial technology policies
    Edited by Jomo K.S. and Greg Felker
24. **Corporatism and Korean Capitalism**
    Edited by Dennis L. McNamara
25. **Japanese Science**
    Samuel Coleman
26. **Capital and Labour in Japan**
    The functions of two factor markets
    Toshiaki Tachibanaki and Atsuhiro Taki

27 **Asia Pacific Dynamism 1550–2000**
Edited by A.J.H. Latham and Heita Kawakatsu

28 **The Political Economy of Development and Environment in Korea**
Jae-Yong Chung and Richard J. Kirkby

29 **Japanese Economics and Economists since 1945**
Edited by Aiko Ikeo

30 **China's Entry into the World Trade Organisation**
Edited by Peter Drysdale and Ligang Song

31 **Hong Kong as an International Financial Centre**
Emergence and development 1945–1965
Catherine R. Schenk

32 **Impediments to Trade in Services**
Measurement and policy implication
Edited by Christoper Findlay and Tony Warren

33 **The Japanese Industrial Economy**
Late development and cultural causation
Ian Inkster

34 **China and the Long March to Global Trade**
The accession of China to the World Trade Organization
Edited by Alan S. Alexandroff, Sylvia Ostry and Rafael Gomez

35 **Capitalist Development and Economism in East Asia**
The rise of Hong Kong, Singapore, Taiwan, and South Korea
Kui-Wai Li

36 **Women and Work in Globalizing Asia**
Edited by Dong-Sook S. Gills and Nicola Piper

37 **Financial Markets and Policies in East Asia**
Gordon de Brouwer

38 **Developmentalism and Dependency in Southeast Asia**
The case of the automotive industry
Jason P. Abbott

39 **Law and Labour Market Regulation in East Asia**
Edited by Sean Cooney, Tim Lindsey, Richard Mitchell and Ying Zhu

40 **The Economy of the Philippines**
Elites, inequalities and economic restructuring
Peter Krinks

41 **Private Enterprise in China**
Edited by Ross Gamaut and Ligang Song

42 **The Vietnamese Economy**
Awakening the dormant dragon
Edited by Binh Tran-Nam and Chi Do Pham

43 **Restructuring Korea Inc.**
Jang-Sup Shin and Ha-Joon Chang

44 **Development and Structural Change in the Asia-Pacific**
Globalising miracles or end of a model?
Edited by Martin Andersson and Christer Gunnarsson

45 **State Collaboration and Development Strategies in China**
Alexius Pereira

46 **Capital and Knowledge in Asia**
Changing power relations
Edited by Heidi Dahles and Otto van den Muijzenberg

47 **Southeast Asian Paper Tigers?**
From miracle to debacle and beyond
Edited by Jomo K.S.

48 **Manufacturing Competitiveness in Asia**
How internationally competitive national firms and industries developed in East Asia
Edited by Jomo K.S.

Manufacturing Competitiveness in Asia

How internationally competitive national firms and industries developed in East Asia

Edited by Jomo K. S.

RoutledgeCurzon
Taylor & Francis Group
LONDON AND NEW YORK

First published 2003
by RoutledgeCurzon
11 New Fetter Lane, London EC4P 4EE

Simultaneously published in the USA and Canada
by RoutledgeCurzon
29 West 35th Street, New York, NY 10001

RoutledgeCurzon is an imprint of the Taylor & Francis Group

Typeset in Goudy by
Keystroke, Jacaranda Lodge, Wolverhampton
Printed and bound in Great Britain by
Antony Rowe Ltd, Chippenham, Wiltshire

British Library Cataloguing in Publication Data
A catalogue record for this book is available from the British Library

Library of Congress Cataloging in Publication Data
Manufacturing competitiveness in Asia: how internationally competitive national firms and industries developed in East Asia/[edited by] Jomo K. S.
p. cm. — (RoutledgeCurzon studies in the growth economies of Asia; 47)
Includes bibliographical references and index.
1. Manufacturing industries—East Asia. 2. Electronic industries—East Asia. 3. International business enterprises—East Asia. 4. East Asia—Foreign economic relations. 5. Competition, International. I. Jomo K. S. (Jomo Kwame Sundaram) II. Series.

HD9736.E18 M36 2003
338.095—dc21 2002036911

ISBN 0–415–29922–5

Contents

Figures and tables

Figures

Tables

Contributors

Rokiah Alavi (PhD, East Anglia) is Associate Professor in the Kulliyah of Economics and Management Sciences, International Islamic University (Malaysia), Kuala Lumpur, Malaysia.

Ha-Joon Chang (PhD, Cambridge) is Lecturer in the Faculty of Economics and Politics, University of Cambridge.

Jaya Gopal (PhD, Imperial College) is a Researcher at the Palm Oil Research Institute of Malaysia (PORIM) in Bangi, Selangor, Malaysia.

Jomo K.S. (PhD, Harvard) is Professor in the Faculty of Economics and Administration, University of Malaya, Kuala Lumpur, Malaysia.

Medhi Krongkaew (PhD, Michigan State University) is Professor of Economics at the National Institute of Development Administration (NIDA) in Bangkok, Thailand.

Park Eul Yong (PhD, Harvard) is Professor and Vice President of Handong University, Seoul, South Korea.

Patcharee Siroros (PhD, Wisconsin) is Associate Professor with the Faculty of Political Science, Thammasat University, Bangkok, Thailand.

Rajah Rasiah (PhD, Cambridge) is Professor and Senior Research Fellow at the United Nations University, Maastricht, Netherlands.

Ken Togo (PhD, Yale) is Associate Professor in the Faculty of Economics, Musashi University, Tokyo, Japan, and was previously with the Foundation for Advanced Studies in Development, Tokyo.

Tseng Ming-Shen (PhD, Wayne State) is Director, Research Division III, Taiwan Institute of Economic Research, Taipei, Taiwan.

Wilaiwan Wannitikul (PhD, Pennsylvania) is Associate Professor in the Faculty of Economics, Thammasat University, Bangkok, Thailand.

Wu Rong-I (PhD, Louvain) is President of the Taiwan Institute of Economic Research, Taipei, Taiwan.

Acknowledgements

On 2 December 1997, the Foundation for Advanced Studies on International Development (FASID) organised FASID Forum III on 'Asia's Development Experiences: How Internationally Competitive National Manufacturing Firms Have Developed in Asia' at the Institute for International Cooperation, Japan International Cooperation Agency (JICA), in Tokyo. This was the culmination of a research project that FASID organised from 1996 to 1998 in partnership with researchers from Taiwan, Korea, Malaysia, Thailand, Singapore, Indonesia and Japan. This volume contains a selection of the research papers presented, which have been substantially revised for the purpose of this publication. I would like to express our appreciation to the numerous research partners, commentators, audience, staff and FASID who made this volume possible. Special thanks go to Kaori Ishii and Mayumi Sugihara from FASID for providing us continued support for the publication of this volume.

Also included is a paper by Ha-Joon Chang of the University of Cambridge, originally for a project initiated by Joseph Stiglitz and managed by Shahid Yusof. Most of the papers were included in a heavily edited volume entitled *Rethinking the East Asian Miracle* edited by Stiglitz and Yusof (Oxford University Press, 2001). Chang's chapter nicely complements the country studies from Korea, Taiwan, Malaysia and Thailand by providing an analytical framework for understanding industrial policy in the East Asian region.

Finally, I wish to thank Brian Folk for his help in preparing this volume for publication.

Jomo K.S.
Kuala Lumpur
August 2002

Abbreviations

ADB	Asian Development Bank
AFTA	ASEAN Free Trade Area
AMD	Advanced Micro Devices
APEC	Asia Pacific Economic Cooperation
ASEAN	Association of Southeast Asian Nations
AT&T	American Telephone and Telegraph
BOI	Board of Investment (Thailand)
BoT	Bank of Thailand
BOT	Board of Trade (Thailand)
CAD	computer-aided design
CAM	computer-aided manufacturing
CBU	completely built-up
CC	continuous casting
CDC	China Development Corporation
CEPT	common effective preferential tariff
cif	cost, insurance, freight
CNC	computer numeric control
CPKO	crude palm kernel oil
CPO	crude palm oil
CPUs	central processing units
CTO	Chief Technology Officer
DEP	Department of Export Promotion (Thailand)
DIT	Department of Industrial Technology (Taiwan)
DFI	direct foreign investment
EA	*Emerging Asia* (published by Asian Development Bank, 1997)
EAM	*East Asian Miracle* (published by World Bank, 1993)
EO	export-oriented
EON	Edaran Otomobil Nasional ([Malaysian] National Car Distributor)
EPZs	export processing zones
ERP	effective rate of protection
ERSO	Electronics Research and Service Organisation
EU	European Union
FASID	Foundation for Advanced Studies on International Development

FDI	foreign direct investment
fob	free on board
FTI	Federation of Thai Industries
FTZs	free trade zones
GATT	General Agreement on Tariffs and Trade
GDCF	gross domestic capital formation
GDP	gross domestic product
GNP	gross national product
GSP	Generalised System of Preferences
HCR	hot charge and rolling
HICOM	Heavy Industry Corporation of Malaysia
HIPASS	hot charge integrated process adjusting and scheduling system
HPAEs	high-performing Asian economies
HSIP	Hsin-Chu Science-based Industrial Park (Taiwan)
IBM	International Business Machines
ICs	integrated circuits
ICA	Industrial Co-ordination Act (Malaysia)
ICT	Industrial Corporation of Thailand
IDMA	International Diamond Manufacturers Association
IFC	International Finance Corporation (World Bank)
IFCT	Industrial Finance Corporation of Thailand
III	Institute for Information Industry
IMC	Industrial Management Company (Thailand)
IMP	Industrial Master Plan (Malaysia)
IS	import substituting
IT	information technology
ITRI	Industrial Technology Research Institute (Taiwan)
JICA	Japan International Cooperation Agency
JPPCC	Joint Public-Private Co-operation Committee (Thailand)
KAIST	Korea Advanced Institute of Science and Technology
KLSE	Kuala Lumpur Stock Exchange
KP	Kwangyang plant (South Korea)
KUB	Koperasi Usaha Bersatu (Malaysia)
LAN	local area network
LCD	liquid crystal display
LLN	Lembaga Letrik Negara ([Malaysian] National Electricity Board)
LMCP	local material content policy
LMW	licensed manufacturing warehouse
LRAC	long-run average cost
MFA	Multi-Fibre Arrangement
MIC	Market Intelligence Centre
MIDA	Malaysian Industrial Development Authority
MIER	Malaysian Institute of Economic Research
MITI	Ministry of International Trade and Industry (Japan)
MNCs	multinational companies

MOEA	Ministry of Economic Affairs
MOU	memorandum of understanding
MPOPC	Malaysian Palm Oil Promotion Council
MSE	minimum scale efficiency
NEP	New Economic Policy (Malaysia)
NESDB	National Economic and Social Development Board (Thailand)
NIC	newly industrialized country
NIE	newly industrialized economy
NSC	National Science Council (Taiwan)
ODM	original design manufacturing
OEM	original equipment manufacturing
PABX	private automatic branch exchange
PBB	Permodalan Bersatu Berhad (Malaysia)
PBE	park-based enterprise
PC	personal computer
PCO	progress control office
PDC	Penang Development Corporation (Malaysia)
PLC	Programmable Logic Controller
PORIM	Palm Oil Research Institute, Malaysia
PORLA	Palm Oil Registration and Licensing Authority (Malaysia)
POSCO	Pohang Steel Corporation (South Korea)
POSTEC	Pohang Institute of Science and Technology
PPO	processed palm oil
Proton	Perusahaan Otomobil Nasional Berhad (Malaysian National Automobile Enterprise)
PVC	polyvinyl chloride
QCC	quality control circle
QES	Quality Evaluation System
R&D	research and development
RCA	Radio Corporation of America
RIST	Research Institute of Industrial Science and Technology
SCADA	supervisory control and data acquisition
SCG	Siam Cement Group (Thailand)
SDI	structurally-depressed industries
SME	small and medium-sized enterprise
SOE	state-owned enterprise
SS	shipping system
STB	Sapura Telecommunications Berhad (Malaysia)
STM	Syarikat Telekom Malaysia (Malaysian Telecommunications Company)
SYCS	slab yard control system
TBA	Thai Bankers Association
TDMA	Thai Diamond Manufacturers Association
TFP	total factor productivity
TGJTA	Thai Gems and Jewellery Traders Association

TMCP	thermo-mechanical controlled process
TSMC	Taiwan Semiconductor Manufacturing Company
TTA	technology transfer agreement
UK	United Kingdom
UMC	United Microelectronics Company
UMNO	United Malays National Organisation (Malaysia)
UNIDO	United Nations Industrial Development Organisation
US(A)	United States (of America)
UTB	Uniphone Telecommunications Berhad (Malaysia)
VAT	value-added tax
VDP	vendor development programme
VISC	Vanguard International Semiconductor Corporation
VLSIC	very large-scale integrated circuit
WTO	World Trade Organisation

1 Introduction

Jomo K.S. with Ken Togo

There are many competing explanations of the East Asian economic miracle, ranging from the cultural to the conjunctural. Even economic explanations are far from being unanimous, with the debate largely over the role of the state and its consequences. There has been considerable debate about the role and nature of the state, and particularly about the consequences of industrial policy. In this debate, there have been three, sometimes distinct, sometimes overlapping, explanations of the role of the state in what the World Bank (1993) has called the East Asian economic miracle which may be summed up as minimalist, market friendly and developmentalist.

The first, essentially *laissez-faire* approach arguing for a minimal role for the state, basically asserts that the state has been largely irrelevant or, even worse, actually obstructive of the essentially market forces which have contributed to rapid growth and structural transformation, including industrialisation. The original and most articulate exponents of this view include Little, Scitovsky and Scott (1970), but there are many supporters of this view. Interestingly, these include the many liberals and neo-liberals who have opposed the Park Jung Hi and subsequent military regimes in South Korea and many 'native' Taiwanese who used to resent suggestions that the mainland Guomindang regime may have contributed to development on that island.

Such a view became especially influential in the early 1980s as the ideological pendulum in the Anglophone world swung to the far right after the election of Mrs Thatcher and Mr Reagan. Intellectually, this swing was bolstered by Keynesianism's apparent responsibility for the fiscal crises and 'stagflation' of the 1970s, the resurgence of monetarism, the emergence of supply-side economics, the public choice school's critique of self-seeking politicians and bureaucrats as well as the property rights school's critique of ill-defined or weak rights as well as greater attention to principal-agent problems. Such views were reflected in what John Toye (1987) has called the 'counter-revolution' against development economics – led by Peter Bauer and Deepak Lal (see references in Toye, 1987), reflected for example in the World Bank's World Development Reports of the early 1980s.

The second, currently popular case for the market-friendly state (World Bank, 1991) was greatly enhanced by the World Bank's (1993) *The East Asian*

Miracle (EAM) study, and is likely to be seen as drawing additional support from the Asian Development Bank's (ADB) (1997) study entitled *Emerging Asia* (EA). Drawing from neo-classical welfare economics, this view accepts the case for government intervention due to the existence and greater significance of externalities and market failures. This approach has given new life to and justification for development economics – which had come under near fatal assault in the early 1980s – by emphasising the more pervasive and deep-rooted nature of externalities and market failures of various types in developing economies. The persistence of such externalities and market failures made the case for what the World Bank (1993) refers to as 'functional' interventions – as opposed to 'market-unfriendly' 'strategic' interventions, which the World Bank did not approve of.

While largely accepting the arguments for state interventions to address market failures, the advocates of the developmental state perspective emphasise that the nature of government interventions in East Asia generally went well beyond the market-friendly functional interventions approved of by the World Bank. While the World Bank disapproved of so-called strategic interventions, the proponents of the developmental state perspective insist that selective industrial policies – involving trade, financial and other interventions – have accounted for 'late industrialisation' in East Asia (Amsden, 1989; Wade, 1990; Chang, 1994).

The key argument is that such interventions have been crucial for developing new industrial capabilities which did not previously exist and which would not have spontaneously emerged due to market forces alone. Thus, the old 'infant industry' argument was resuscitated, with insights from Gerschenkron's (1962) observations on the advantages of economic 'backwardness' as well as the requirements of 'late industrialisation'. The developmental state advocates emphasised the role of 'strong states' (in Myrdal's sense (1968)) as well as the manipulation, if not distortion, of market mechanisms to achieve developmental objectives. 'Market-enhancing' (Aoki *et al.*, 1997) and other critiques of the earlier emphasis on wasteful rent-seeking behaviour has shown how contingent rents have served as incentives for achieving such goals which go well beyond the neo-classical welfare economics notion of market failures. There has also been greater appreciation of co-operative and associational solutions to co-ordination failure and other collective action problems.

There is, of course, considerable variation in perspectives within the three camps, as well as positions which may be seen as intermediate. For example, a significant number of institutionalists have identified and emphasised collective action problems and co-ordination failures, which may be best addressed by direct government intervention or, alternatively, by private sector collective initiatives, or by improved government–private sector consultation, or even by corporatist institutions and mechanisms. In so far as some such problems may not be generally acknowledged as market failures, the related solutions may not be seen as within the pale of acceptable market-friendly interventions. And in so far as the intervention may be anticipatory or pro-active, rather than reactive, it is more likely to be seen as strategic rather than functional.

As noted earlier, the World Bank's *East Asian Miracle* (1993) approves of market-friendly functionalist interventions – such as ensuring good governance, sound macro-economic management, physical and social infrastructure provision and high savings and investment rates – while eschewing market-distorting strategic interventions. Nevertheless, given the significance of the latter, particularly in Northeast Asia, the *East Asian Miracle* study considered the impact of strategic interventions, particularly 'directed credit' and 'industrial policy-related trade interventions'. The *East Asian Miracle* study insisted that the latter failed in East Asia, while conceding that 'directed credit' worked. However, the World Bank suggested that the conditions and circumstances of such limited success in Northeast Asia were very unusual, if not unique (Confucianism, bureaucratic capability, favourable initial and international conditions, etc.), and therefore not to be emulated.

Growth accounting exercises – suggesting little total factor productivity (TFP) growth in most of the region – have also been invoked by the World Bank, Paul Krugman (1994) and others to suggest the inferiority of East Asian growth in achieving technical progress. The main conclusion drawn is that rapid growth in the region has largely been due to massive factor (capital and labour) inputs due to high savings and investment rates, foreign direct investment, growth of the wage labour force in the formal sector and human capital investments. Further factor inputs are bound to run up against diminishing returns, and rapid East Asian growth cannot be sustained, at least at the breakneck pace of the past three decades.

Many East Asians were deeply offended by Krugman's (1994) comparison of East Asian growth with that of the Soviet Union in earlier times, and the implications that East Asian economic performance has not been all that miraculous and that slower growth is unavoidable and imminent. However, there has been less critical attention to the bases of his analysis, namely the more conventional neo-classical growth accounting exercises by Alwyn Young (1994) on the one hand, and the more heterodox exercise by Kim Jong-Il and Lawrence Lau (1994).

This is not the place to go into an extended discussion of the theoretical as well as methodological issues involved. However, Dani Rodrik (1994, 1995) observes (see also Collins and Bosworth, 1996) that while 'the evidence on investment rates is direct and speaks for itself, the evidence on TFP is indirect and has to be interpreted with care'. Also, more recent findings (Collins and Bosworth, 1996; Bosworth and Collins, 2000) suggest that East Asian economies have been evolving toward greater TFP gains since the 1980s as they attain higher stages of development. They also argue that future growth in the region can be sustained as the educational and skill profiles of the labour forces continue to grow. There is also greater appreciation of the crucial conceptual differences and inter-relationship between TFP and cost competitiveness.

Krugman is probably right in claiming that the new endogenous growth theory cannot be invoked against his arguments as even higher TFP residuals would then be expected. However, if technological learning only becomes important beyond a certain stage of development or when technological progress requires changes in the labour process more conducive to such learning and shop-floor innovation, one would have different expectations of TFP growth in East Asia outside of Japan.

But even if we accept the theoretical and methodological bases for Krugman's claims (which are not unproblematic), there is good reason to suspect his conclusion of lack of technological progress when one considers the consequences of differences in price determination in different product markets which affect growth accounting exercises. In this case, the important distinction is between the more technologically sophisticated products, enjoying legally protected monopolistic rents, and other more mass-produced products in far more competitive markets. The differences in the nature of the labour markets have also had some bearing on product price determination. Most East Asian workers outside of Japan and, perhaps, Singapore have been under-remunerated owing to international labour immobility, among other factors, resulting in the relative under-pricing – and hence competitiveness – of East Asian exports in international trade.

The different economic performances of the three regions considered by *Emerging Asia* (ADB, 1997) do not merely involve differences in economic growth, or even of structural transformation, though these are not unimportant. Before the 1990s, the World Bank's first-tier East Asian high-performing Asian economies (HPAEs) (including Singapore) grew by almost two percentage points more than the three second-tier Southeast Asian newly industrialising countries (NICs) (Malaysia, Thailand and Indonesia); the difference was even greater on a per capita basis owing to the higher population growth rates in the latter. When one considers the far larger contribution of natural resource rents to the growth performance of these three NICs, the achievement of the HPAEs is even greater.

Whereas the *East Asian Miracle* study obscured this difference, the *Emerging Asia* study addresses it in terms of regional differences. Unfortunately, neither study pays sufficient attention to the major policy differences between the two regions and their consequences in terms of 'late industrialisation'. Industrial policy has been far more extensively and effectively deployed in Japan, South Korea and Taiwan than in the second-tier Southeast Asian NICs. The success of such industrial policy is reflected in the greater industrial and technological capabilities of the former compared to the latter.

Neither study comes to terms with the fact that Japan, South Korea and Taiwan selectively kept out foreign direct investment (FDI), with FDI only accounting for a modest share of gross domestic capital formation (GDCF), whereas FDI has been far more important in Southeast Asia, especially in Singapore and Malaysia, and that too partly for political reasons. Both studies also repeat the neo-liberal mantra of trade liberalisation and economic openness without fully acknowledging the critical difference between 'free trade' à la Little *et al.* (1970) and the 'simulated free trade' juxtaposition of export subsidies against import protection à la Bhagwati (1986, 1988) – as in Northeast Asia.

In making regional generalisations, the *Emerging Asia* study glosses over many important differences within the three main regions considered. In reviewing the *East Asian Miracle* study, Dwight Perkins (1994) suggested that generalisations about East Asia obscured the existence of at least three distinct East Asian types among the eight HPAEs – the Northeast Asian HPAEs (including Taiwan), the Southeast Asian HPAEs and the two city states of Hong Kong and Singapore. The significance of industrial and technology policies as well as state-owned enterprises in the island republic, in contrast to the recently returned British colony, underscores the difficulties in making facile generalisations. Any alternative categorisation would also be moot, but the recognition of such variety is often obscured in stressing regional similarities. State-owned enterprises have performed well in Singapore and perhaps in Taiwan as well, but less well in Malaysia and Indonesia, which is not surprising given the circumstances of their establishment and management.

Competition, openness and exports?

The ADB's *Emerging Asia* (1997) study argued that market competition, openness and export orientation were the key ingredients of East Asia's miraculous economic performance. It is not possible to refute these claims comprehensively here, but fortunately others have already done so very persuasively.

On the claim of market competition, one can refer to the World Bank's (1993) discussion of the importance of 'contests' in East Asia. Consistent with the Austrian School critique of the neo-classical economic fetish for perfect competition, East Asian governments have not been insistent on competition to avoid wasteful, excessive competition and to enable firms to achieve economies of scale. Contests or managed competition as well as managed exposure to international markets have instead been used to force firms to become internationally competitive as quickly and as reasonably as possible.

As Bhagwati (1986, 1988) and many others have noted, the East Asian governments have not been open to free trade, as suggested by the *Emerging Asia* study. Bhagwati has argued that free trade has been 'simulated', with import protection in East Asia offset by export subsidies, but this is certainly not free trade as normally understood. Nor were East Asian governments all open to FDI as suggested by the *Emerging Asia* study. FDI in Japan, South Korea, and even Taiwan has accounted for a smaller proportion of gross domestic capital formation than is the norm for developing countries. Even in the Southeast Asian HPAEs, all with higher than average FDI/GDCF, there has been significant regulation of FDI.

The *Emerging Asia* study also ignores the problems of liberalisation and openness, such as the causes and consequences of the 1997 financial crisis in Southeast Asia. Contrary to the claim that 'the market' will exact swift and painful punishment on governments and economies that do not have their macro-economic house in order, the timing, nature and consequences of the 1997 financial crisis in Southeast Asia underline the imperfect nature of financial markets, as reflected in the long delay in 'rectification'. In a world economy where foreign exchange spot transactions are worth more than seventy times total international merchandise trade transactions, the financial sector has become increasingly divorced from the real economy. With the recent proliferation of new financial instruments and markets, the financial sector has an even greater potential to inflict damage on the real economy.

Even George Soros (1997) has argued that the unregulated expansion of capitalism, especially finance capital, threatens to undermine the system's viability and future, i.e. that capitalism has to be saved from itself. While admitting that he himself has profited greatly from financial liberalisation, he argues that excessive liberalisation has resulted in virtual anarchy, which is dangerous for the stability so necessary for the orderly capitalist growth and democratic development desired by his liberal vision of a Popperian 'open society'.

Ever since Lord Keynes advocated 'throwing sand' into the financial system to check the potentially disastrous consequences of unfettered liberalisation, Keynesians – and others – have been wary of the financial liberalisation advocated by ideological neo-liberals and their often naïve allies. Nobel laureate James Tobin has called for a tax on foreign exchange spot transactions to enable more independent national monetary policy, discourage speculative capital movements and increase the relative weight of long-term economic fundamentals against more short-termist and speculative considerations, besides more than adequately funding the United Nations system and programmes. As many have pointed out, the international financial system and its further liberalisation have favoured those already dominant and privileged in the world economy, largely at the expense of the real economy and development in the South.

Dani Rodrik (1994) has challenged the *East Asian Miracle* study's claim of the significance of export orientation. The economic histories of Japan, South Korea and Taiwan suggest that most industries began by producing for the domestic market as has been typical of import-substituting industrialisation. The East Asian difference has been in effectively requiring and facilitating the rapid transition to production for export, often through the creative deployment of trade policy. For instance, effective protection has often been provided by some East Asian governments for limited periods of time conditional upon export promotion, i.e. the export of products protected for sale in the domestic market; such a contingent rent ensured that infant industries quickly became internationally competitive – both in terms of cost as well as quality – rather than manufacturing dwarfs producing only for the protected domestic market.

All this is not to imply that industrial policy has always been well motivated and successfully deployed. The World Bank's claim of trade policy failure is methodologically problematic, and does not even bother to distinguish government interventions motivated by different considerations, e.g. the desire to enrich a politically influential, or otherwise favoured concessionaire. The *Emerging Asia* study cites problems with the Korean heavy and chemical industrialisation drive, but just as with the policy failures attributed to the Japanese Ministry of International Trade and Industry (MITI), such selective evidence is not conclusive proof of the inevitable failure of industrial policy in principle.

The *Emerging Asia* study is quite correct in emphasising the new constraints in the articulation, elaboration and implementation of industrial policy, especially those imposed by the new international economic governance, particularly through the World Trade Organisation (WTO). But instead of urging Asian governments to work together in their common interest to resist the emerging international economic governance, the *Emerging Asia* study urges precisely the opposite.

Unfortunately, neither the *East Asian Miracle* nor *Emerging Asia* studies go very far in trying to explain or understand why government interventions have, on balance, accelerated structural transformation and resulted in the development of significant industrial and technological capabilities in East Asia, and to a lesser extent, in Southeast Asia. This suggests that better understanding of the institutional basis and consequences of government intervention can take us some way towards greater appreciation of some reasons for the different outcomes of government intervention in the three main Asian regions considered by the *Emerging Asia* study.

This study, however, proceeds quite differently. By studying how industries and firms in the East Asian region became internationally competitive, we highlight the micro-economic determinants of their economic performance as well as the relevant institutional features which made rapid growth, structural change and industrial progress possible. This will necessarily entail greater appreciation of the role of industrial policy, in the sense of selective pro-active government interventions, as well as other institutional features, such as technological learning and information sharing arrangements, as well as other initiatives to overcome collective action problems.

The various East Asian country – or more accurately, industry and firm – studies show how particular firms and industries were shaped by and, in turn, reshaped environmental factors, including macro-economic and meso-economic conditions, in order to achieve international competitiveness. Such international comparison highlights the importance of national conditions, including the role of government policy. While such studies are not new for Japan and, to a lesser extent, for the first-generation newly industrialised economies (NIEs), there are almost no such studies for the second-tier Southeast Asian newly industrialising countries. Studies of successful companies in Southeast Asia have mainly focused on diversified conglomerates, offering little insight on how specific industries have become internationally competitive.

Unlike Northeast Asia, most export-oriented manufacturing production in Southeast Asia is owned or controlled by transnational companies locating plants in the region to lower production (especially labour) costs, evade environmental or other restrictive regulations (e.g. for use of ozone-layer diminishing CFCs) or to access particular markets (e.g. under the MFA or Multi-Fibre Arrangement). Consequently, there has been very little attention paid to the determinants of international competitiveness for nationally-owned or domestically-controlled firms and industries, the primary focus of this volume.

The differences in national industrial capabilities between Japan and the first-tier NIEs on the one hand and the second-generation Southeast Asian NICs on the other are arguably also due to the different nature and roles of government. It is generally agreed that the quality of government intervention has been superior in the former, and is now reflected in superior industrial capabilities as well as greater international competitiveness in technologically more sophisticated production.

With the moot exception of Singapore, it would now be generally acknowledged that government intervention throughout East Asia has been wracked by varying degrees and types of rent-seeking. However, rent-seeking seems to have had a stronger influence on policy-making in Southeast Asia (barring Singapore) than in Northeast Asia, where rent-seeking has tended to take advantage of policies and regulations established for other purposes. Hence, while there may have been some abuse, the concept of contingent rents has been more successfully elaborated and implemented in Japan and the first-tier NIEs compared to the Southeast Asian NICs.

Successful rent-seeking has not necessarily led to excessive dissipation of accumulated rents. Whether or not rent-seeking necessarily leads to the dissipation or waste of rents depends a great deal on the structure or conditions of rent creation and capture. In so far as rents have been specifically created as instruments of selective intervention, autonomous and strong states have been able to ensure that they be effectively deployed to achieve desired policy goals, including the rapid development of industrial capabilities. Hence, for example, temporary effective protection conditional on export promotion has provided an attractive incentive to induce firms to invest in rapidly improving industrial and technological capabilities in order to be internationally competitive. Also, successful rent capture has been reflected in higher rates of accumulation, savings and investment throughout the East Asian region.

In the country chapters which follow, relatively little attention is given to what the World Bank calls functional interventions. Instead, in trying to understand the conditions favourable to the successful development of particular industries, the emphasis is necessarily on so-called strategic or selective interventions. This is not to deny the significance and contribution of functional interventions, many of which have been important, if not crucial, to the success of particular industries. However, this is a more familiar story on which much has been written (e.g. World Bank, 1993).

The focus in this volume on the more selective interventions is inevitable given the micro- and meso-perspective of firms and industries. This also highlights the case for industrial policy and selective interventions, recognising the specific nature of some market failures and other adverse conditions which need to be successfully addressed and overcome for particular firms and industries to become internationally competitive. Also, in a situation of scarce fiscal and other government resources as well as the limited range of feasible and efficient industrial policy instruments available, selectivity also makes sense – especially for states with some capabilities and knowledge – as well as strategic sense for the medium and long term. While government powers, regulations and interventions can be abused by policy-makers and implementers as well as by others seeking to take advantage of such policies, this is true of all government roles, and is hardly a case specifically against selective interventions.

Industrial policy success in East Asia

Chapter 2 by Ha-Joon Chang begins with a critical review of the debate over selective government support for sectoral growth or industrial policy. He shows that there are more theoretical justifications for industrial policy than acknowledged by the World Bank (1993), and that some of these justifications (e.g. co-ordination of competing investments, scale economies) were probably more important in the actual formation of industrial policy in the East Asian countries than the ones acknowledged by the Bank (e.g. 'Big Push', formation of implicit cartels in international negotiations).

Chang argues that the true effects of industrial policy are still poorly understood owing to methodological inadequacies. He also challenges the view that other countries wanting to adopt East-Asian-style industrial policy should not do so because they do not have the necessary capabilities. First, industrial policy does not really require an exceptionally competent bureaucracy. Second, bureaucratic capabilities develop through deliberate efforts and with experience. Third, he notes that more market-oriented systems also require high institutional capabilities.

While acknowledging the new constraints on industrial policy associated with recent economic liberalisation, especially the advent of the WTO, Chang emphasises the remaining options for industrial policy manoeuvring. He also argues that the recent East Asian crises do not prove that industrial policy has ultimately been detrimental as those most adversely affected were those which had abandoned industrial policy. Chang also notes that while economic maturity, democratisation and the rise of private sector power may limit industrial policy options, they do not render industrial policy unfeasible.

The remaining chapters in this volume go beyond debates at the macroeconomic level to focus attention on how domestic or indigenous firms and industries in East Asia became and remained internationally competitive. We focus here on the development of manufacturing industry, widely acknowledged as the engine of economic growth. We also presume that the technologies necessary for long-term economic development are mostly acquired through learning by doing. Our case studies shed light on various aspects of manufacturing development at either the industry or the firm level.

As Park shows in Chapter 3, the establishment of South Korea's Pohang Iron and Steel Company, Ltd (POSCO) – a huge integrated steel mill and now the world's largest steel producing company – was not supported by most economists when proposed in the 1960s. POSCO required substantial investment and future performance was uncertain. The South Korean government, however, defied such criticisms and established POSCO as a state-owned enterprise (SOE). It used the government budget, foreign aid and other loans for investment, and utilised technology transferred from more advanced countries, mainly Japan. Now, POSCO records high profits and supplies cheap steel to domestic manufacturing industries like the auto industry.

Park suggests that government intervention, management autonomy, rapid learning and technological innovation were the important elements for achieving this result. He uses the concept of 'social capability' to explain the success of POSCO, emphasising how certain social systems and other features of a society make scarce factors more productive. His study suggests the importance of dividing feasible industrial interventions and policy instruments in line with known state and societal economic capabilities. Park's analysis also implies a more realistic view of technological innovation.

In Chapter 4, Wu and Tseng analyse the role of government in the development of Taiwan's information industry, a high-tech industry. The Taiwan government established the Industrial Technology Research Institute (ITRI) in the Hsin-Chu area in 1973 to repatriate Taiwanese engineers who had been working in the United States in high-tech industries by providing them with attractive research conditions. In 1980, the government established Hsin-Chu Science-based Industrial Park (HSIP) in the same area to provide high-tech ventures with good infrastructure, tax incentives and financial support. These two institutions helped Taiwanese high-tech engineers spin off from ITRI to start high-tech ventures in HSIP. Today, HSIP is the centre of gravity for Taiwan's information industry. A third of Taiwan's portable computers, one-fifth of desktop computers, 30 per cent of motherboards, and over 70 per cent of colour scanners are produced in the Park. This case study stresses the importance of complementarity between institutions in overcoming market failures and achieving developmental objectives. Wu and Tseng also emphasise how different modest selective interventions were developed for different segments of the information technology (IT) industry at different points in time.[1]

In Chapter 5, the Malaysian case studies, which include an electronics components supplier, a government-favoured telecommunications firm, the national car industry and palm oil refining, had very different experiences as well. Eng Hardware, the case study from the electronics industry in Malaysia, started business in 1976 as a backyard workshop engaged in repair work and simple metal fabrication, and grew to become a high-precision machine tool factory supplying services to multinational companies (MNCs). Jomo *et al.* suggest that a major element contributing to its growth has been 'trust' among ethnic Chinese, linking the founder of Eng Hardware, the director of Intel (who decided to subcontract to Eng Hardware) and the Penang state government, which encouraged such linkages and co-ordinated meetings between local firms and MNCs. Trust among Eng Hardware, Intel and the Penang state government facilitated co-operation among them, with Intel providing technical training and finance to Eng Hardware for the improvement of its competitiveness. This case study highlights the potential as well as limitations of manufacturing to supply multinational firms.

The Sapura telecommunications group has utilised its close relationship with the Malaysian government to expand its business, e.g. by supplying telephone sets to the Telecommunication Department for handsome profits. Sapura has used its profits to invest in research and development, besides buying foreign licences and technical assistance from Siemens and Bell telephones. Nowadays, Sapura exports some of its products to industrialised countries such as Singapore, the Netherlands, Germany and Japan. However, the availability of attractive profits has caused Sapura to invest in the automotive industry as more intense competition has made telecommunications less attractive. The government emphasis on inter-ethnic economic redistribution may well have compromised the efficacy of Malaysian industrial policy.

The third Malaysian case study looks at the chequered performance of one of the showpieces of Malaysia's state-led industrialisation: the Proton national car project. State-created rents were an important part of the incentive structure that allowed technologically backward Malaysia to hope to enter the global car market. The Proton story seems to be one of limited successes with major challenges ahead. Although more than a decade old, Proton's financial viability still depends on large rents created for it through the protection of the domestic market. The authors argue that progress towards the technology frontier has been slow, which indicates ineffective co-ordination and monitoring by the state, poor incentives for performance improvements and perhaps even the weak bargaining position of the Malaysian state *vis-à-vis* its Japanese technology suppliers. Also, the political need to transfer some rents to relatively inefficient Malay parts suppliers could be an important part of the problem.

Palm oil refining in Malaysia highlights the limitations of industrial recommendations based on static comparative advantage analysis. In the early 1970s, palm oil refining in Malaysia was judged economically unprofitable. The government, however, challenged this view and imposed export duties on crude palm oil, allowing varying levels of export duty exemption on processed palm oil. Such rents for palm oil refining encouraged many firms to invest in it. With severe competition due to excessive investments, oil-refining firms rapidly improved their efficiency and became a major supplier of processed palm oil in the world.

For Thailand, in Chapter 6, the gems and jewellery industry and the Siam Cement Group were selected as case studies of the development of manufacturing industry. Patcharee and Wilaiwan with Medhi describe the industry's development in detail, suggesting that the industry association has been a major element in the growth of the gems and jewellery industry. The small producers in the industry organised a business association called the Thai Gems and Jewellery Traders Association (TGJTA) which influenced the government to promote this industry. The government lifted import duties and business taxes on rough gem-stones in 1977, while the business tax on finished jewellery manufactured for export was also lifted in 1981. Such policy changes have clearly benefited the growth of the industry, and further government support may be necessary for the industry – that may have reached a plateau in its development – to progress.

The Siam Cement Group started business in 1913 as a royal company that supplied cement for construction, including physical infrastructure, in the country. The Siam Cement Group has enjoyed huge profits under generous protection from the government, expanding the scope of its business into industries that use cement as raw material as well as industries not related to cement. The authors argue that the pattern of Siam Cement's investments suggests forward and backward integration, while industrial diversification into activities highly protected by the government has contributed to the continued success of the company. They also mention the decision-making system, leadership, corporate culture, connections, research, human resource development, and long-term strategy as other factors contributing to its success.

From an institutional perspective, these case studies can be categorised into three groups. The first group shows that close ties with government have clearly contributed to manufacturing sector growth, as in the cases of POSCO in South Korea, Taiwan's information technology industry, the Siam Cement Group in Thailand, palm oil refining and Sapura in Malaysia and the ship repair/building industry in Singapore (Wong, 2001b). Although the types of intervention by the governments were quite different in each case, these studies offer good evidence of pro-active industrial policy. The challenge for this group is to analyse how efficiency can be sustained and enhanced with government intervention.

The second group shows the importance of linkages among indigenous firms in developing countries and MNCs. This group includes Taiwan's information technology industry, Eng Hardware in Malaysia and the electronics industry in Singapore (Wong, 2001b). Although government played an important role in each case by facilitating contact, the main role of the government has been catalytic. The challenge for this group is to analyse how technology transfer was successfully carried out to mutual advantage.

The third group reflects the importance of private institutions that have emerged to address collective action problems. This group includes the gems and jewellery industry in Thailand and the raw silk industry in Japan (Togo, 1998). Business associations have mainly been organised by private initiatives and we need to better understand why private institutions have been successfully organised and sustained in certain places and not in others.

Any comparative study of this type is hard pressed to draw generic lessons of likely use and relevance to others seeking to emulate them. It is important to emphasise something simple which should be quite obvious by now, namely that export-oriented labour-intensive manufacturing did and does not develop spontaneously with the availability of cheap labour, free trade and the absence of capital controls on foreign investors. Besides various functional interventions, other supportive conditions (e.g. cultural and linguistic affinities, law and order) and policies (e.g. incentives including tax breaks and subsidies, specialised education and training, investment and export promotion) have often been decisive for inducing desired investments.

While it has become quite fashionable to speak of East Asian models or to think of the rest of East Asia as clones or imitators of the Japanese, the case studies in this volume clearly show the variety to be found in East Asia. The contrast between Japan and the first-tier NIEs on the one hand and the second-tier Southeast Asian NICs on the other needs to be reiterated. While industrial policy efforts have successfully supported the emergence of many internationally competitive domestically-owned industries and firms in the former, Southeast Asian progress has been far less impressive.

Owing to market imperfections arising from economies of scale, uncertainties or both, companies find it easier not to compete in international markets if they can enjoy highly profitable domestic sales. It is therefore socially desirable for the government to impose export targets in return for protection in the domestic market. Such provision of effective protection conditional on export promotion appears to have been a critical difference between Northeast Asian late industrialisation and other – including Southeast Asian – experiences of import-substituting infant industry protection, with the infant industry never really growing up to compete internationally.

The more modest Southeast Asian achievement in this regard is, of course, due to several factors, including the weakness of industrial policy-making and implementation as well as the greater reliance on FDI to develop export-oriented manufacturing capacity. Although Singapore has relied even more on FDI to accelerate progress on technological learning curves and to achieve greater domestic value addition, it has used non-trade industrial policy instruments as well as state-owned enterprises far more extensively and successfully.

This variety is not only sub-regional or national, or industry-specific or even firm-specific, but also historical and conjunctural, in the sense of firms and industries responding to particular circumstances and opportunities. The accumulated evidence offered in this volume makes it difficult to claim that industrial organisation or structure, management style or organisational form, firm size or even cheap labour, 'initial conditions', cultural factors or social networks, or any other single factor or combination of factors has contributed to the rise of East Asian firms or industries.

While macro-economic stability, good governance, adequate provision of physical and social infrastructure, high savings and investment rates as well as other such factors must certainly have created conditions favourable to successful development of firms and industries, the main common factor emerging from the case studies appears to be firm or industrial capacity for rapid and successful technological learning and innovation. The detailed case studies have emphasised the largely evolutionary, incremental and cumulative dimensions of such technical change, much of which can only be meaningfully understood in the context of the industry or firm, once again underlining the case for selective interventions. Owing to market imperfections in the supply of much information on technology, gains are likely to be particularly large as there are likely to be economies of scale in the acquisition and dissemination of such information.

Since companies that invest in training fear being unable to recoup their costs, training is likely to be under-funded without state co-ordination and support. In many developing countries much more can be done on the education and training front to strengthen human resource development conducive to rapid industrialisation. Governments should play a major role in providing technical and vocational training as well as the relevant secondary and tertiary education to prepare personnel for accelerated industrial development. The government can influence and co-ordinate the supply of and demand for specific skills and should flexibly anticipate and effectively co-ordinate preparation of human resource requirements in the medium and long term.

Governments, however, have an unfortunate tendency to overemphasise formal education while neglecting the significance of actual work experience and training on the job. Hence, incentives should also be offered to encourage in-house training, but when firms are reluctant to make such investments, e.g. for fear of other firms 'free-riding' on them, it is often necessary for the government to step in. The evidence summarised in this volume emphasises the dialectical or interactive nature and experiential basis for the generation of new knowledge and practices. These learning and innovation capabilities seem to have developed best when firms and industries have been subjected to effective but not stifling discipline – whether by the market or by the state (e.g. hard budget constraints, high penalties for failure, policy discipline and flexibility, including allowing or even requiring exit for failure) – and been induced by government leadership (to overcome co-ordination and other collective action problems) and support (i.e. industrial policy) as well as the prospect of extraordinary profits (i.e. rents).

The prospect of capturing more rents has often been crucial in motivating improved industrial practices and thus, technological change, as recognised by the Schumpeterian perspective. This has often led to an emphasis on strengthening intellectual property rights to support the monopolistic claim to rent for the innovator. However, much of the technological innovation discussed in the case studies has involved 'learning by doing', with little of it easily codifiable as the basis for property rights claims. Instead, trust, loyalty, co-operation and other social relations violating the arm's length norm favoured by market ideology seem to have been very important in explaining the rapid technological learning characteristic of most internationally competitive industries and firms in East Asia.

Owing to market imperfections due to risk and uncertainty, firms may under-invest in long-term production facilities since the rate of profit required by the companies is likely to be too high. Hence, long-term investment is likely to be smaller than 'socially desired' unless the state helps to induce it. While foreign investments, borrowings and aid can augment investments, in most circumstances national savings are the primary determinant of investments over the long run. As long as they are positive, the actual level of real interest rates does not seem to be a major determinant of savings and investment rates. Instead, macro-economic stability and rapid income growth seem to be more important.

Tax and other incentive policies allowing firms to capture and retain more of such rents has been reflected in high retained earnings, which not only serve as a strong incentive for rapid technological learning, but also for high company savings as well as reinvestment of profits. This regime has been crucial for the rapid rate of accumulation throughout the region. High firm savings have been crucial to the region's high savings rates since mandatory provident fund schemes have only raised household savings rates in Malaysia and Singapore.

When well conceived for industrial policy, rather than for other rentier purposes, deliberate market distortions are shown by many of our case studies to have been crucial trade policy instruments which have encouraged the emergence of internationally competitive industries. In other words, our case studies underline that industrial policy works best within the context of well-conceived development strategy driven by strong political commitment. Such political will is rarely found when the horizons of the political leadership are primarily short-term and largely dictated by compromise among strong vested interests, as in the Thai political system. The conventional emphasis on the choice between market and state as alternative means for resource allocation is also shown to be wanting. Neo-liberals will also be disappointed that public ownership has not stood in the way of achieving international competitiveness as seen in the Singaporean and South Korean cases.

A government's supportive role should be ongoing and not only limited to helping industries and firms achieve international competitiveness. However, such an ongoing role must be flexible and appropriate, capable of adjusting to and changing with new circumstances, particularly to address new problems of market as well as state failures and constantly changing conditions. Market failures are usually defined and understood in a static neo-classical sense, but the inability of markets to spontaneously bring about desirable structural transformations, e.g. in building dynamic comparative advantage, is ultimately the most important reason for industrial policy. While state intervention may not always be a superior solution to a market failure, a state failure may be better addressed by different government intervention rather than by relying on the market.

Hence, as most of the studies in this volume show, it is not the extent or degree of government support which counts, but rather its nature and appropriateness. In the cases of Japan and Thailand, where government intervention appears to have been far less important in the past, at least for the industries studied in this volume, institutional innovations to overcome collective action problems seem to have been crucial. The Thai authors are concerned that the absence of greater government support will stand in the way of continued progress of the Thai gems industry.

All this is not meant to imply that government interventions in East Asia have all been an unqualified success. There are many well-known examples of such failure in East Asia as elsewhere. Though the nature of the studies of this volume do not allow us to say too much about this matter, it is important to recognise that not all government interventions in East Asia have been motivated by industrial policy objectives. Many if not most of the well-known failures of government intervention have in fact had little to do with industrial policy as such. And the industrial policy failures in Malaysia and Indonesia since the 1980s are clearly cases of ill-advised ambitions of the politically powerful, even opposed by economists favouring industrial policy in principle.

Acknowledgements

We are grateful to all who offered useful critical feedback on the first draft of this introduction, and are particularly appreciative of Professor John Toye's challenging queries.

Note

1 An important comparative counterpart is the case of Singapore, which has been analysed extensively in the work of Wong Poh-Kam and his colleagues (see Wong 2001a, b), originally prepared for the FASID project upon which this book draws heavily. Their work highlights Singapore's main development policies in response to the changing political-economic context and external environment: while still heavily dependent on foreign MNCs, Singapore has also developed indigenous manufacturing capabilities in certain sectors with policies and programmes to assist local enterprises. Singapore's industrial policy has focused largely on infrastructural support, human resource development and promoting investments in high value-adding manufacturing activities, mainly by foreign firms. Their two contrasting case studies of the electronics and ship repair/building industries illustrate some of the variety of possible policy stances available to governments seeking to promote particular industries. Singapore's industrial policy has thus contributed significantly to the development of the country's manufacturing capabilities, offering interesting lessons for other developing countries constrained by few natural resources and a small domestic market.

References

Amsden, Alice (1989) *Asia's Next Giant*, New York: Oxford University Press.

Aoki, M., H.-K. Kim and M. Okuno-Fujiwara (eds) (1997) *The Role of Government in East Asian Economic Development: Comparative Institutional Analysis*, Oxford: Clarendon Press.

Asian Development Bank (ADB) (1997) *Emerging Asia*, Manila: Asian Development Bank.

Bhagwati, Jagdish (1986) 'Rethinking trade strategy', in J. Lewis and V. Kallab (eds) *Development Strategies Reconsidered*, New Brunswick, NJ: Transaction Books.

—— (1988) 'Export-promoting trade strategy: issues and evidence', *World Bank Research Observer*, 3(1), January: 27–57.

Chang Ha-Joon (1994) *The Political Economy of Industrial Policy*, Basingstoke: Macmillan.

Collins, Susan M. and Barry P. Bosworth (1996) 'Economic growth in East Asia: accumulation versus assimilation', *Brookings Papers on Economic Activity*, 2: 135–203.

Gerschenkron, Alexander (1962) *Economic Backwardness in Historical Perspective*, Cambridge, MA: Harvard University Press.

Kim Jong-Il and Lawrence Lau (1994) 'The sources of economic growth of the East Asian newly industrialized countries', *Journal of the Japanese and International Economies*, 8(3): 235–71.

Krugman, Paul (1994) 'The myth of Asia's miracle', *Foreign Affairs*, 73(6): 62–78.

Little, Ian M.D. (1981) 'The experience and causes of rapid labour-intensive development in Korea, Taiwan Province, Hong Kong and Singapore, and the possibilities of emulation', in Eddy Lee (ed.) *Export-led Industrialisation and Development*, Asian Employment Program, Geneva: International Labour Organisation.

—— (1994) 'Trade and industrialisation revisited', *Pakistan Development Review*, 33(4.i), Winter: 359–89.

Little, I., T. Scitovsky and M. Scott (1970) *Industry and Trade in Some Developing Countries: A Comparative Study*, New York: Basic Books.

Myrdal, Gunnar (1968) *Asian Drama*, 3 vols, New York: Pantheon.

Perkins, Dwight (1994) 'There are at least three models of East Asian development', *World Development*, 22(4): 655–61.

Rodrik, Dani (1994) 'Getting interventions right: how South Korea and Taiwan grew rich', Working Paper no. 4964, Cambridge, MA: National Bureau of Economic Research.

—— (1995) 'Trade strategy, investment and exports: another look at East Asia', Working Paper no. 5339, Cambridge, MA: National Bureau of Economic Research.

Soros, George (1997) 'The capitalist threat', *The Atlantic Monthly*, February.

Togo Ken (1998) 'Elements of the Japanese raw silk industry: an explanation by the NK model', in Takahashi Kazuo (ed.) *Asia's Development Experiences*, Tokyo: Foundation for Advanced Studies an International Development: 40–74.

Toye, John (1987) *The Dilemma of Development*, Oxford: Blackwell.

Wade, Robert (1990) *Governing the Market*, Princeton: Princeton University Press.

Wong Poh-Kam (2001a) 'The role of the state in Singapore's economic development', in P.K. Wong and C.Y. Ng (eds) *Industrial Policy, Innovation and Economic Growth: The Experience of Japan and the Asian NIEs*, Singapore: Singapore University Press.

—— (2001b) 'Singapore: electronics and shipbuilding/repair industry case studies', in P.K. Wong and C.Y. Ng (eds) *Industrial Policy, Innovation and Economic Growth: The Experience of Japan and the Asian NIEs*, Singapore: Singapore University Press.

World Bank (1981) *World Development Report 1981*, New York: Oxford University Press.

—— (1983) *World Development Report 1983*, New York: Oxford University Press.

—— (1991) *World Development Report 1991*, New York: Oxford University Press.

—— (1993) *The East Asian Miracle: Economic Growth and Public Policy*, New York: Oxford University Press.

—— (1997) *World Development Report 1997: The State in a Changing World*, New York: Oxford University Press.

Young, Alwyn (1994) 'Lessons from the East Asian NICs: a contrarian view', *European Economic Review*, 38(3–4): 964–73.

—— (1995) 'The tyranny of numbers: confronting the statistical realities of the East Asian growth experience', *Quarterly Journal of Economics*, 110(3): 641–80.

2 Industrial policy and East Asia

The miracle, the crisis and the future

Ha-Joon Chang

The issue of industrial policy has, arguably, been at the heart of the debate on the East Asian developmental experience during the past two decades or so. In the late 1970s and the early 1980s, there was an intense international debate on the issue, largely prompted by Japanese industrial success during the first three decades of the post-war period.[1] In the late 1980s, similar policies in the next tier of East Asian success stories such as South Korea (henceforth Korea) and Taiwan stirred up the second phase of the debate.

Unlike in the case of Japan, where the interventionist nature of industrial policy was widely (if not universally) acknowledged, the orthodoxy regarding Korea and Taiwan until the early 1980s was that they were free-market, free-trade economies with little industrial policy (e.g. Ranis and Fei, 1975; Balassa, 1982). From the early 1980s, however, there emerged a number of research findings that questioned this orthodoxy and emphasised the role of industrial policy in the economic success of these countries (Jones and Sakong, 1980; Evans and Alizadeh, 1984; Amsden, 1985; Luedde-Neurath, 1986). Partly as a consequence of these studies, some mainstream commentators started to acknowledge the existence of industrial policy in these countries, although they insisted that it had not made much of a positive impact. The alleged failure of industrial policy measures in these countries was supposedly because they were self-cancelling (e.g. the 'virtual free trade' position of Little, 1982; and World Bank, 1987) or because they were ostensibly porous (e.g. the theory of 'proscriptive vs. prescriptive' intervention proposed by Bhagwati, 1988). The publication of the works by Amsden (1989) on Korea and Wade (1990) on Taiwan were the culmination of the so-called 'revisionist' offensive that had started in the early 1980s, to which the World Bank's 1993 *The East Asian Miracle* (EAM) report was the mainstream's most conciliatory answer.[2]

Contrary to the expectation of its authors, the *EAM* report failed to put an end to the debate. First of all, important methodological and empirical criticisms of the report were made (see the special symposium in *World Development*, 1994 vol. 22, no. 4; Fishlow *et al.*, 1994; Singh, 1994), to which its authors have not provided convincing answers. Second, some issues were, in my view, inadequately addressed, both by the authors of the report and its critics, in the earlier debate surrounding the report. A more balanced assessment of the role of industrial policy in East Asia requires examination of these issues. Third, the recession in Japan and the 1997–8 financial crises in a number of other East Asian economies, which have occurred since the publication of the *EAM* study, have made popular the view that industrial policy created economic problems, rather than miracles, in the region. Given that one major conclusion of the *EAM* was that industrial policy had few positive impacts, rather than that it was harmful, these recent events call for a re-examination of the role of industrial policy in the region.

This chapter is organised in the following manner. First, a critical review of the *EAM* is presented. More effort will go into raising issues neglected in the earlier debate, rather than to going over issues already debated. Next, the currently popular view that industrial policy was behind the 'downfall' of the East Asian model will be assessed. Whether recent economic, political, and institutional changes (both at national and international levels) have made the use of industrial policy in East Asia less feasible in the future is considered next. This is followed by concluding remarks.

Industrial policy in *The East Asian Miracle* report: contributions and limitations

The *EAM* report distinguished itself from previous publications by the World Bank and most mainstream economists on the role of industrial policy in East Asia in at least two respects. First of all, it acknowledged the existence of industrial policy in most non-Japanese East Asian countries. In the case of Japan, the existence of interventionist industrial and trade policies had been widely acknowledged earlier, while the very existence of industrial policy was a matter of intense debate in the case of Korea and Taiwan even until the late 1980s.[3] The *EAM* accepted the contention of the 'revisionists' that the extent and degree of industrial policy in these countries were much larger than what mainstream economists had cared to admit before, and started its discussion from there.

Second, the *EAM* clearly accepted a number of important theoretical justifications for industrial policy – such as the so-called 'Big Push' argument and the existence of learning externalities (see next section). This was a big contrast to many earlier mainstream works, which argued that market failures were limited to areas such as infrastructure, education, and health, and that there was therefore no reason for governments to intervene in industry, especially at the sectoral level.

Having abandoned the earlier mainstream practice of dismissing the issue of industrial policy as theoretically unjustifiable and/or largely absent from East Asia, the *EAM* resorted to two more practical arguments to come up with its negative verdict on industrial policy. First of all, it tried to show empirically that despite its widespread existence, industrial policy had not made much difference either to the production structure or productivity performance of East Asian countries. Second, it argued that, whatever its contribution to the development of some East Asian countries may have been, industrial policy cannot be adopted by developing countries today because they face different domestic and international conditions.

It was argued that the latter countries lack the domestic institutions needed for effective implementation of East-Asian-style industrial policy (especially a competent bureaucracy). Also, the kind of 'permissive' international trading environment that the East Asian countries enjoyed during the time when they actively used such policy (that is, between the 1950s and the 1970s) has ceased to exist, especially following the conclusion of the Uruguay Round of General Agreement on Tariffs and Trade (GATT) negotiations.

In the rest of this section, three aspects of the *EAM*'s verdict on industrial policy mentioned above are critically examined to highlight some issues inadequately dealt with in the earlier debate. First, the *EAM*'s (partial) theoretical acceptance of industrial policy, next, its empirical refutation of the success of industrial policy in East Asia, and finally, practical objections it raises to the transferability of industrial policy to other countries.

Theoretical justifications for industrial policy

The *EAM* acknowledged some important justifications for industrial policy, unlike earlier orthodox publications on the subject (see World Bank, 1993: 90–3, 293–4). First, the need to co-ordinate complementary investments, in the presence of significant scale economies and capital market imperfections, was acknowledged – this is the well-known 'Big Push' argument. Second, the role that a state can play in organising domestic firms into implicit cartels for negotiating with foreign firms or governments was recognised. Third, the importance of learning externalities was emphasised.

However, almost in the same breath, the *EAM* dismissed another important theoretical justification for industrial policy, namely, infant industry promotion, on the grounds that its success is not guaranteed. This refutation is rather peculiar, since all other theoretical justifications for industrial policy accepted by the *EAM* also cannot guarantee the success of policies based on them. But apart from this rather obvious point, there is not much added value to be gained from reiterating the theoretical arguments already accepted by the *EAM*. Instead, a few other theoretical justifications for industrial policy more or less ignored by the *EAM* (and, in fact, by many of its critics as well) and their implications will be discussed below.

Co-ordination of competing investments

The first under-explored justification for industrial policy is the need to co-ordinate investments, not simply complementary investment projects but also competing projects, to 'manage competition'. This was actually the central point of contention in the industrial policy debate of the early 1980s over the Japanese experience, but was curiously ignored by the *EAM*.

The logic here is that the oligopolistic competition that characterises many modern industries with significant scale economies often leads to excess capacity, unless there is co-ordination of investment activities among competing firms.

Excess capacity leads to price wars, which damage profits for the firms concerned, which may force them to scrap some of their assets, or even lead to bankruptcy.

Needless to say, asset scrapping and bankruptcy are useful and cost-less ways of rearranging property rights in a world without transaction costs and 'specific assets' (Williamson, 1985), but we do *not* live in such a world. This means that the specific assets involved also have to be scrapped or re-allocated to alternative uses that can create much less value from the assets concerned, thus incurring social waste. If the emergence of excess capacity can be prevented through *ex ante* co-ordination of competing investments, such social waste may be reduced (for more detailed arguments, see Chang, 1994, ch. 3; also see Telser, 1987; and Amsden and Singh, 1994).

Many mainstream economists have argued that excess capacity is a non-issue, especially for small economies that are price-takers, because what cannot be consumed in the domestic market can always be exported. However, this is often not a viable option, at least in the short run (and it is the short run that counts here). First, at least since the late 1970s, many industries have been suffering from chronic over-capacity at the world level.[4] Moreover, real-world markets are often segmented along the lines of quality, design, and geography, and therefore the 'world market' may not be as big as it seems, since it takes time and resources to break into new market segments. In addition, some small economies have deliberately built capacities well beyond their domestic markets and have become price-makers, rather than price-takers, even on the world scale. For example, despite being a relatively small economy, Korea is the world's first or second largest producer of ships (depending on the year) and the third largest producer of microchips (the largest for memory chips only) and, therefore, what the country produces does have an important impact on world prices. Indeed, this is why the end to the earlier practice of co-ordination among competing investments became such a problem in Korea (see next section for more details).

Given these considerations, there is a clear theoretical justification for co-ordinating competing investments. And, indeed, such co-ordination has been one of the most important components in the industrial policy regimes of the East Asian countries. This has been manifested in their continuous concern for 'excessive competition' or 'wasteful competition' and attempts to minimise redundant investments through mechanisms such as industrial licensing and investment cartels (see Chang, 1993, for further details).[5] By ignoring this important issue, the EAM ended up neglecting a central component of industrial policy in East Asia.

Further implications of scale economies

The EAM certainly recognised the importance of scale economies in modern industrial development, when it discussed co-ordination of complementary investments. However, this is not the only way in which scale economies matter. First, scale economies have important implications for the cost competitiveness of a country's industries. In industries with significant scale economies, choosing a sub-optimal scale of capacity can often mean 30 to 50 per cent differences in unit costs. For this reason, the East Asian governments have used measures such as industrial licensing, government procurement, export requirements, and subsidies to ensure that factories are built at scales which are not too much below (and hopefully above) the minimum efficient scale. Of course, this has invited criticisms on anti-trust grounds, but the official attitude has been that monopolistic firms producing at optimal scale are much less of a drag to the economy than 'competitive' firms all producing at sub-optimal scales.

Second, scale economies also has a hitherto-ignored link with luxury consumption control (for a more detailed discussion, see Chang, 1997a). The well-known practice of luxury consumption control in East Asian countries – most notoriously, but by no means exclusively or even mainly, through import control – has often been interpreted as no more than a thinly disguised protectionist ploy. Alternatively, it has been seen as a manifestation of the paternalistic desire to impose what the government sees as a 'sound consumption pattern' (the phrase was explicitly used in, for example, the Fourth Five-Year Plan document of Korea, p. 27). However, there has been much more to these controls. First of all, it was thought important to restrict conspicuous consumption for the purpose of reducing class conflicts, especially given the (real and imagined) communist threat. Second, there was the desire to maximise the investible surplus by repressing luxury consumption out of profit. Third, and most relevant to our discussion here, restrictions on the consumption of luxury varieties in industries like the passenger car industry, where consumer demand for variety is important, were regarded as important for enabling producers to attain the maximum possible scale in production.[6]

To sum up, while the *EAM* acknowledged the crucial role of scale economies in necessitating the co-ordination of complementary investments, it did not explore their role beyond this. However, while it may sound less fancy than co-ordinating complementary investments and giving industrial development a 'big push', ensuring the achievement of scale economies in key industries was, in practice, probably a much more important aspect of East Asian industrial policy than the former.

'Protective' industrial policy, social insurance and structural change

Another aspect of industrial policy that has received little recognition in the East Asian context is its 'protective' role. It is widely believed that what distinguishes industrial policy in East Asia is that it concentrated on 'picking winners', rather than 'protecting losers', as was often the case in other countries. There is certainly a large element of truth in this view. However, protective industrial policies were also widespread in East Asia, if less so than in other countries. Therefore, we need to go deeper if we are to understand why protective industrial policy in East Asia did not end up blocking structural change, as in many other countries.

We argue that protective industrial policies in East Asia served two functions. The first was the more short-term one of providing 'social insurance' to firms which are in temporary difficulty, but cannot borrow their way out due to capital market imperfections.[7] Like the policy of co-ordinating competing investments, the practice can be justified in terms of asset specificity as it will be socially inefficient to scrap specific assets in the face of a temporary setback. This would be the case if the net present values of future income streams from assets are larger than the cost of the support needed to keep them employed in their current uses (assuming full asset specificity – namely, their value in alternative uses is zero). The best example of such policy is the famous Japanese practice of sanctioning (but closely supervising and disciplining) 'recession cartels' in industries deemed to be in temporary difficulty (see Dore, 1986; and Chang, 1994, ch. 3, for more details).

The second, and probably more important, function was the more long-term one of promoting structural change. When an industry is in need of a large-scale adjustment, those who have made specific (human and physical, or even relational) investments in the industry face a situation where their next best option is totally scrapping their assets and thus drastically reducing their incomes. Unless there is a mechanism that allows them an acceptable level of income during the transition period, when they run down their existing assets and re-tool themselves (e.g. purchase new equipment, re-train workers), they will have an incentive to resist change by political means. In such a situation, measures to reduce the impacts of adjustment on the owners of specific assets can accelerate, rather than slowdown, structural change by reducing political resistance to change, if they also provide incentives for (physical and mental) re-tooling (for a more detailed argument, see Chang and Rowthorn, 1995).[8]

In Japan, 'cartels for structurally-depressed industries' (or SDI cartels) were granted to declining industries in return for their efforts to phase out obsolete capacities and upgrade technologies (Dore, 1986, provides a fascinating study of this experience; also see Renshaw, 1986). During the late 1980s, some declining industries in Korea, such as textiles, received temporary supports (e.g. subsidies for equipment upgrading, exemption from anti-trust law). This was done through rationalisation programmes sanctioned by the Industrial Development Law, on the condition that they achieved certain targets in relation to technology upgrading (see Chang, 1993, for details).

What distinguishes these policies from similar policies in other countries is that they were 'forward looking' in the sense that they made it explicit that the aim of the protection was not to preserve the industries concerned, but to phase them out 'in an orderly manner' or to technologically upgrade them. Perhaps more importantly, they also had well-specified performance targets for the beneficiaries, thus preventing the policies from becoming 'nursing homes' for declining industries. In other words, because of the way they were designed and implemented, protective industrial policies in East Asia seem to have promoted, rather than hindered, structural change.

To summarise, by concentrating on 'developmental' industrial policy, the *EAM* ignored 'protective' industrial policy. Such industrial policy is often regarded as blocking structural change, and therefore not justifiable, but it has actually played a positive role in East Asia in two ways. First, it has provided social insurance to producers experiencing a temporary difficulty but who cannot borrow their way out of it due to capital market imperfections. Second, and more importantly, it has promoted structural change by easing the difficulties involved in moving 'specific' resources out of the declining sectors or in upgrading them. Such policy was probably not the most important aspect of East Asian industrial policy, but was by no means unimportant, especially for the more advanced countries like Japan and Korea.

Empirical refutation

At the risk of some simplification, the essence of the *EAM*'s empirical verdict on the role of industrial policy in East Asia can be summarised as follows: there is no evidence that the industries promoted by industrial policy had higher output growth or more rapid productivity growth than other industries.

The methodologies and data used in the study have already been subject to a range of well-known criticisms, including the problems inherent in the definition and the measurement of total factor productivity (see Lall, 1994; Kwon, 1994; Rodrik, 1994; and Singh, 1994, among others; also see Chang, 1995, appendix). As they are mostly of a technical (which, of course, does not mean 'trivial') nature, the summary of which may take up considerable space, this section does no more than raise a couple of methodological points that have not been adequately considered in the earlier debate on the *EAM* study.

The *EAM* tested the effects of industrial policy by trying to correlate the extent of government support for an industry (however measured) with the industry's performance. A major problem with this approach is that one major justification for industrial policy lies in the existence of externalities. Hence, it is, by definition, very difficult (if not entirely impossible) to measure its effects at the sectoral level (the 2-digit level in this case), as its effects will spill over into other sectors. And, if we can measure the effects of such policy, we probably would not need it in the first place. Since externalities are difficult to measure, it is therefore difficult to quantify the effects of industrial policy measures intended to internalise externalities. Therefore, if the effects of an industrial policy are easy to quantify, it is unlikely that the externalities involved were significant.

The *EAM* does acknowledge this problem, but justifies its sectoral approach on the grounds that spill-over effects are mostly confined to 'closely related sectors, often sectors that would be identified with a two-digit classification' (p. 326). The problem with this conclusion is that it is based on one study on the pattern of spill-overs of research and development (R&D) in industrial economies. Apart from the danger of drawing such a strong conclusion from a single study, it is not clear how relevant such a study is to understanding the role of industrial policy in developing economies. For developing economies, where R&D plays a minor role at best, the more important spill-over effects may include the formation of a skilled labour force and the increase in generalised engineering capability. Therefore, the result of the study on R&D spill-over cited by the *EAM* only has limited relevance, even if it were true (for more details, see Chang, 1995, appendix).

The empirical study presented in the *EAM* also suffers from a serious 'identification problem'. Lall (1994) has already pointed out that the *EAM* classifies industries at the 2-digit level, which is too general a level of classification to identify the activities promoted – typically, industrial promotion has been targeted at a much more focused level, sometimes even involving support defined at the firm level. However, such detailed classification may also be too fine for some purposes, because some major components of industrial policy, such as export promotion, is usually conducted at a much broader, cross-sectoral level.[9] However, there is a more fundamental element in this 'identification problem', which is that the *EAM* did not bother to find out which industries, at whatever levels of classification, were *actually* promoted.

The statistical work conducted in the *EAM* study is based on the presumption that the East Asian governments promoted industries that had higher value-added or higher capital-intensity. However, the problem is that the choice of industries to be promoted in these countries was never made on simple criteria like 'capital intensity' or 'value-addition'. Rather, the choice often reflected a whole set of considerations, including, to name just a few, international market conditions, availability of relevant domestic technological capabilities, and the net foreign exchange implications of promoting the industry concerned. For example, the Korean textile industry, which the *EAM* regards as the quintessential 'non-promoted' industry (p. 316), in fact received a lot of promotional support even *after* the government launched its heavy and chemical industrialisation programme in 1973 – it even had a special promotional law in 1979. This was because of the industry's critical role as the main supplier of foreign exchange (it was the largest export industry until well into the 1980s), which was necessary for the country to import capital equipment and buy technology licences needed for 'infant' industries. Hence, the fact that the Korean textile industry was, according to the EAM, unusually large by international standards is *not* proof of the failure of Korean industrial policy, as the *EAM* argues, but rather of its success (for more details, see Chang, 1995, appendix).

In other words, the *EAM* has classified industries into those promoted and those that were not, according to what its authors *thought* was the industrial policy practice in the East Asian countries, rather than according to what *was* the practice in these countries. Such disregard for what was actually going on in the countries concerned is quite similar to its failure to theoretically consider (even if only to disapprove) the justifications for some central components of industrial policy in these countries. These included co-ordination of competing investments, policy measures to attain scale economies, and the use of 'protective' industrial policy.[10]

The replicability problem

Very importantly, the *EAM* argues that industrial policy requires certain conditions to be successful. Therefore, other countries that do not meet such conditions cannot hope to use such policy successfully. Two kinds of arguments were made in this regard. First, it was argued that, in order to make industrial policy work, even to the (allegedly) limited extent that it worked in East Asia, a country needs certain institutions. In this connection, it is generally implied if not asserted that a highly capable bureaucracy was essential and can only be found in the East Asian countries of Japan, Korea, Taiwan, Hong Kong and Singapore. Second, it has been argued that industrial policy is not feasible any more because the new international economic regime that emerged out of the Uruguay Round of the GATT talks has made 'illegal' the industrial policy tools used by the East Asian countries. How persuasive are these arguments?

Institutional capability

The *EAM* argued that successful management of industrial policy, as one of those 'selective' or 'strategic intervention' policies that go beyond the 'functional intervention' to address 'market failures', requires certain unusual institutional capabilities that can rarely be found outside East Asia. The report argues that effective organisation of 'contests' among recipients of state support, necessary for successful industrial policy, requires 'the competence, insulation, and relative lack of corruptibility of the public administrations in Japan and Korea' (p. 102).[11] The report then concluded that the more market-oriented economies of Southeast Asia provide a better role model for emulation by other developing countries. Their success, it is claimed, proved that there is a lot that countries with poor administrative and other institutional capabilities can derive from concentrating on the 'fundamentals', which does not require advanced institutional capabilities (macroeconomic stability, human resource development, agricultural development, among others).[12]

The problem with this argument is not that anyone seriously doubts that effective conduct of selective industrial policy (or for that matter, any policy) requires a bureaucracy that has the competence and political influence to impose 'hard budget constraints' on the recipients of state support according to relatively transparent rules. This proposition is, in fact, what many 'revisionists' have repeatedly emphasised; so, at that level, there is really no dispute. The problem is that the *EAM* implicitly assumed that the more 'selective' a policy is, the more difficult it is to administer, and thus the more institutional 'props' (such as a good bureaucracy) it needs – or, to put it differently, the closer an economic system is to the *laissez-faire* ideal, the easier it is to run it. Is this true?

First, well-functioning markets require institutional prerequisites as much as well-functioning policies require them, although they may be somewhat different – developed contract law, an efficient capital market, and an effective dispute settlement mechanism, to name just a few – because, without these institutions, market exchange becomes very costly (Chang, 1997b). A successful modern free-market economy will also require highly capable *private sector bureaucracies* that can successfully manage large and complex firms. The enormous difficulties that many developing and transition economies are having in constructing the basic institutions of a market economy and private sector bureaucratic capabilities are clear testimonies to the fact that more market-oriented economic systems are *not necessarily* easier to construct and run than more interventionist systems.

Second, in the same vein, it is not clear at all whether industrial policy necessarily requires a more capable bureaucracy than the so-called fundamental policies. This will depend on the nature and scale of the policies concerned. For example, is promoting a small number of relatively unsophisticated industries necessarily more difficult than, say, administering a large-scale primary educational programme? For another example, running good macroeconomic policy in the face of a large (positive or negative) external shock is often a lot more difficult than running selective industrial policy (as the East Asian economies found to their chagrin in the 1990s). The point is *not* that industrial policy is necessarily more (or less) difficult to run than other policies, but that one cannot make a categorical statement about the ease or the difficulty of a particular type of policy without looking at the particular issues involved.

Third, it is not clear whether capable bureaucracies in East Asia were the products of 'highly unusual historical and institutional circumstances' (p. 366). At first sight, this seems more than reasonable. We all know that the East Asian countries have had at least a thousand (more in the case of Chinese-speaking countries) years of meritocratic bureaucracy, and this surely must prove that these countries are highly unusual – or does it?

Let us answer this question by first thinking about Singapore. Is it really the Confucian tradition that has made its bureaucracy what it is? The principles that lie behind the Singaporean bureaucracy seem more British than Confucian. Take the case of Taiwan. When its bureaucracy was running mainland China before 1949, it already had the longest tradition of meritocracy and competitive recruitment in the world, but that did not prevent it from being one of the least competent and most corrupt bureaucracies of the time. Did Korea always have an exceptionally competent bureaucracy? The Korean bureaucracy was notorious for its incompetence and nepotism in the 1950s (Cheng *et al.*, 1998), and was sending its bureaucrats for extra training to countries like Pakistan and the Philippines even until the late 1960s. It was only through continuous efforts at civil service reform, and *not* as a result of history and tradition, that Korea managed to create a competent and relatively clean bureaucracy – a point that even the *EAM* acknowledged in passing (World Bank, 1993, Box 4.4).

The point is *not* that history and tradition do not matter, but that capabilities (and the institutions that embody them) can be built and destroyed a lot more easily than is assumed in the *EAM* (and by many others). It is true that capability building often takes time, but this is not the same as saying that countries that do not have high capability should never try 'difficult' policies (such as industrial policy). Such capability can be, and often has been, built rather quickly, not least because there is also 'learning-by-doing' in administration as in production. Institutions are, in other words, subject to imitation and innovation, as are technologies (Westney, 1987). Indeed, the World Bank itself has later come around to accept, although still not wholeheartedly, this line of criticism, as can be seen from its 1997 *World Development Report* that emphasised the need for state capability building (World Bank, 1997).

Changing international trading environment

The *EAM* cites the birth of the new international trading regime, following the conclusion of the Uruguay Round of GATT talks and the creation of the World Trade Organisation (WTO) in 1995, as a severe constraint on the effectiveness of the interventionist trade and industrial policy measures used by the East Asian countries (pp. 25, 365). While it accepts that there is some room for manoeuvre,[13] its verdict on the effect of the WTO regime on developing country policy autonomy seems overly pessimistic.

To begin with, one should not exaggerate the additional constraints on trade and industrial policies that the WTO regime has brought about by talking as if everything was allowed under the pre-Uruguay Round regime. The old regime also had many restrictions on the range of acceptable policy instruments. Therefore, the East Asian countries had to exercise a lot of ingenuity in choosing the industrial policy means and diplomatic skills to iron out problems with their trading partners.

Second, it should not be forgotten that the WTO regime is still an evolving system. The collapse of the Seattle talks in November 1999 shows that there are still some fundamental disagreements among the member countries of the WTO regarding the shape and the running of the new trading system (see Chang and Evans, 2000, for further discussion). Moreover, even if some political consensus on the broad principles of the new trading system can be forged, the exact characteristics of the system will be determined only with the accumulation of precedents over time. As with any other legal system, its principles are stated in fairly general terms, and therefore need to be actively 'interpreted'.

Third, we need to point out that restrictions on the use of subsidies in the WTO regime are not as binding as portrayed by the *EAM* and elsewhere. For instance, there still are subsidies which are perfectly legal (the so-called 'non-actionable' subsidies). These include 'non-specific' subsidies and certain types of 'specific' subsidies (for basic R&D, agriculture, disadvantaged regions, and equipment upgrading to meet higher environmental standards). There are also subsidies which are 'actionable' (e.g. the trading partner can impose countervailing duties), but not prohibited. However, in this case, the complaining country has to prove that the subsidy concerned has caused 'material damage', which is not easy or worthwhile, especially when it concerns developing countries with tiny market shares. The only subsidies that are prohibited outright are subsidies that require recipients to meet certain export targets or to use domestic goods instead of imported goods. However, the poorest countries (roughly defined as countries with less than US$1,000 annual per capita income) are, in fact, exempt from even this.

Fourth, as in the pre-WTO regime, under the WTO regime, countries are allowed to raise tariffs or introduce quantitative restrictions when they have balance of payments problems. Since almost all of them are in a permanent balance of payments crisis, this provides significant room for manoeuvre for the developing countries. Indeed, it was actually almost invariably on this ground, rather than under the infant industry provision of the GATT, that the East Asian countries imposed the tariffs and quantitative restrictions they used for infant industry promotion (Akyuz *et al.*, 1998: 31). Of course, these measures are supposed to be commensurate to the scale of the balance of payments problem, which means that there is a clear restriction on the magnitude of measures that can be used. However, the WTO expressly allows individual countries to choose *where* to impose these measures (i.e. how they define 'non-essential imports'), so there is actually significant room for selectivity in the use of these measures, which is, after, all what the debate is about.

To summarise, it is true that the WTO regime has put greater restrictions than before on the range of acceptable trade and industrial policy tools. However, the restrictions are by no means as wide-ranging and severe as the *EAM* suggests, and there is more room for manoeuvre for developing countries, especially the poorest ones, which are given some special exemptions. Given that the pre-WTO international trading regime was by no means permissive, it seems doubtful whether the new international trading regime makes past industrial policy practices in East Asia as irrelevant for other developing countries as they are said to be.

The East Asian crisis, Japanese recession and industrial policy

The debate on East Asian industrial policy took a new turn following the continued recession in Japan and the economic crises in a number of other countries in the region.[14]

As already pointed out, until recently, many mainstream economists have tried hard to deny the very existence of industrial policy in East Asia. Many mainstream economists who acknowledged its existence (including the authors of the *EAM*) were basically arguing that it made little, if any, difference to the economies concerned. With the economic troubles in the region since the late 1990s, however, many of those who denied the existence or effectiveness of industrial policy in East Asia have made an intellectual U-turn. Instead, they now argue that industrial policy was indeed widespread in the region, and effective in the negative sense of creating inefficiencies and encouraging excessive risk-taking (for a more comprehensive critique of this argument, see Chang, 2000).

Before discussing the role of industrial policy in the East Asian crisis, one needs to put this crisis into perspective. While the scale of crisis in many countries in the region is truly mind-boggling, the whole region is not collapsing. Taiwan is still going strong and Singapore has managed to keep its head above water. As for Japan, the problem seems to be more one of perception rather than reality (although this is not to say that the recession can therefore be ignored). True the country was in the longest recession in its post-war history during the 1990s, but even then, its relative performance *vis-à-vis* the US, which is supposed to have entered a new 'Golden Age', seems quite respectable. For example, if we use the

World Bank data set, its average per capita GDP growth rate between 1990 and 1997 was 1.0 per cent, a rate not much below the 2.0 per cent attained by the US. More recent data are difficult to come by, but combining the GDP growth data for 1990–9 and data on population growth rates (1990–8), we get per capita GDP growth rates of 1.6 per cent and 1.4 per cent for the US and Japan respectively (*Financial Times*, 2000, and World Bank, 1999). According to the *Economist*, between 1989 and 1998, the average per capita GDP growth rates in Japan and the US were in fact *identical* at 1.6 per cent. Meanwhile, the labour productivity growth rate was actually *higher* in Japan at 1.2 per cent, compared to 0.9 per cent for the US (*The Economist*, 10 April 1999: 67). These figures suggest that the current Japanese economic recession by no means marks the demise of its economic system.

The main difficulty with the argument that industrial policy was behind the Asian crisis is that it is in fact mostly the more market-oriented Southeast Asian countries and Hong Kong, rather than the industrial-policy states of East Asia, that are in crisis. In the Japanese case, there is a widespread agreement that the recent economic problem has been caused by poor macroeconomic policy, rather than industrial policy. Despite their industrial policy, Taiwan and Singapore are not experiencing crises. Of course, the inclusion of Korea, well known for its industrial policy, in the list of crisis countries complicates things, but one begins to see a more consistent pattern when noting that Korean industrial policy was actually largely dismantled by the mid-1990s.

To begin with, let us consider the Southeast Asian countries. While the EAM underestimated the role that industrial policy played in these countries – it played an important role in developing some natural-resource-related industries (e.g. see Jomo and Rock, 1998) – it is undeniable that industrial policy was not a major element in their policy regimes. Thailand and Indonesia have had little industrial policy, except for agricultural processing industries in the case of Thailand and a few 'prestige' projects (e.g. aircraft) in the case of Indonesia. Malaysia has had more systematic industrial policy, but it can hardly be described as a critical factor in the country's policy regime in the way that it was in the East Asian countries. Indeed, during the past decade or so, many observers of Southeast Asian countries have argued that the absence of industrial policy is precisely why they were finding it increasingly difficult to upgrade their industry and exports. In short, industrial policy could not have been a major factor behind the crises in the Southeast Asian economies, simply because there was little of it around. Indeed, real estate investments, that had nothing to do with industrial policy, rather than industrial investments were principally responsible for the Southeast Asian bubbles (see Henderson, 1999, for more details).

But was Korea not one of the few archetypal industrial-policy states, and would not its crisis therefore prove the defects of industrial policy? The fact that the over-investments that caused the country's crisis were mostly in industries – rather than in real estate development, as in Southeast Asia – also seems to corroborate this argument. However, contrary to popular perception, industrial policy was largely gone in Korea in the build-up to the crisis. Slowly from the late 1980s but very rapidly from 1993, with the inauguration of the Kim Young Sam administration, the Korean government dismantled industrial policy, except for R&D support in some high-technology industries (see Chang, 1998b, for further details). Therefore, it is difficult to blame the Korean crisis on industrial policy as it no longer existed in any meaningful way.

In fact, one can go even further to argue that it was actually the *demise* of industrial policy, rather than its continuation, that was the major factor behind the current crisis in Korea (see Chang *et al.*, 1998; and Chang, 2000, for further details). Most importantly, the end to the policy of investment co-ordination among competing firms allowed the proliferation of redundant investments in key industries that constituted one major cause of the crisis and is now the subject of the so-called 'Big Deal' industrial restructuring (see note 5 for more details).

To summarise, contrary to popular perception, recent economic problems in East Asia do *not* show that industrial policy was a major drag on the economies of the region. Above all, as there was little industrial policy around in the crisis countries (including Korea that had largely dismantled such policy by the mid-1990s), it seems highly implausible to argue that industrial policy was responsible for the crisis. On the contrary, it can even be argued that it was the absence of such policy that contributed to the crisis, at least in some of the countries concerned.

Some thoughts on the future of industrial policy in East Asia

What is the future for industrial policy in East Asia? To some, this question may sound pointless, given the wide-ranging liberalisation measures instituted following the IMF packages implemented in the region, and also given, at least for the moment, the conversion of most governments in the region to liberalisation.

However, formal laws and rules cannot fully determine the working of an economic system – after all, the 'idiosyncratic' Japanese and German economic systems developed out of the very American formal institutional structures that the Occupation Authorities imposed after the Second World War. Moreover, policy needs and fashions change, and it is therefore not certain whether governments in the region will maintain their current policy stances in the future. Therefore, it is still useful to explore structural trends to consider whether governments in the region can still use activist industrial policy in the future, should their political commitments and objectives change.

Economic maturity

During the past decade or so, it has been popular among researchers of the more advanced East Asian economies (Japan, Korea, and Taiwan) to argue that growing economic maturity of these economies has made industrial policy almost impossible to implement successfully. There are two variants to this argument – one based on the problem of complexity, and the other based on the problem of uncertainty.

The complexity argument is that, with economic development, economies become more complex, and therefore become more difficult to administer centrally. This argument is widely accepted as a truism, but should be critically reconsidered. It is true that, other things being equal, a more complex problem increases the informational requirements for successful policy solution, and is therefore more difficult to manage centrally. But the problem is that other things are *not* equal.

First of all, a more mature economy is likely to have better administrative capability, if only because its bureaucracy will have had more experience of 'learning-by-doing'. As pointed out above, learning-by-doing is not confined to production activities. The implication is that even a relatively 'simple' policy will be difficult to administer for developing country bureaucracies with low capability, while more advanced economies have bureaucracies which can deal with quite complex policies with greater ease.

Second, a more developed economy is typically better organised into larger and better-managed units (e.g. large modern corporations, producer associations, community organisations). This means that it is easier to implement a given policy in a more mature economy, as the latter is likely to have more effective 'intermediate' enforcement mechanisms. Indeed, this was precisely one of the factors that Marx and his followers (including the 'anti-socialist' Schumpeter) thought would make socialism increasingly feasible with economic development. The point can also be made from the opposite end. It is well known that industrial policies are typically very difficult to implement in industries where firms are very small and not organised into industry or regional associations.

In short, a more mature economy typically (if not always) poses more complex challenges, but at the same time typically has better capabilities (both at the governmental and the social level) to manage those tasks. Therefore, it is not clear whether centralised co-ordination through industrial policy necessarily becomes more difficult with economic development and maturity.

A related, but slightly different, line of argument is based on the problem of uncertainty, rather than complexity. The argument is that, when a country reaches the frontier of technological development, it becomes much more uncertain what the government should do to help industry. This argument is more compelling than the one based on complexity just examined.

However, to say that industrial policy becomes more difficult in 'frontier' industries is not to say that this makes industrial policy impossible in a mature economy. For instance, most of the justifications of industrial policy reviewed above should hold for frontier industries too. Indeed, some of these justifications may become even stronger with economic maturity (e.g. learning externalities). Moreover, even in a frontier industry with considerable uncertainty about its future, there is no reason why an intelligent bureaucracy – in close consultation with the private sector – should not be able to identify broad trends and provide support for certain types of productivity-enhancing activities. The best examples of the successful use of industrial policy in frontier industries is provided by the experience of Japan during the 1980s and the early 1990s (see Okimoto, 1989; Fransman, 1990; and Weiss, 1998, for some examples).

Unfortunately this argument, which may be quite relevant in the context of some frontier industries in the most advanced economies, has been inappropriately deployed for other situations. Even in the most advanced countries like Japan, there are many industries still catching up. When it comes to economies like Korea and Taiwan, the argument is unconvincing. Despite what Koreans, for example, like to think, these economies are still a good two to three decades behind Japan in almost all industries. Therefore, if industrial policy worked well in Japan as late as the late 1980s and the early 1990s, it should work for Korea and Taiwan in the early decades of the new millennium, if not necessarily beyond. Needless to say, the argument is basically irrelevant when it comes to the Southeast Asian economies.

To sum up, the first variety of the 'maturity' argument – namely, the complexity argument – is not compelling. As economies mature, policy implementation capability increases, both at governmental and at 'intermediate' levels, and therefore it is not clear whether they necessarily become more difficult to manage centrally. The second variety – namely, the uncertainty argument – is more convincing, but its applicability is limited, and mainly applies only to Japan among the East Asian countries. Moreover, even with overall economic maturity, a country will still have a lot of industries where technological capabilities have not yet reached the world's frontier. And even in those industries at the frontier, the more sensible solution is often not the abandonment of industrial policy, but its appropriate modification, as the Japanese experience since the 1980s suggests.

Democratisation

It has long been argued that interventionist industrial policy requires strong states which can override sectional interests. Therefore, it is argued, the increasing democratisation of many East Asian countries during the past decade or so should make such policies politically less acceptable and therefore less feasible. This argument is especially used in relation to countries like Korea and Taiwan, which have recently gone through substantial democratisation.

However, it is not clear whether industrial policy is incompatible with a democratic polity. Countries like France, Japan, Austria, Norway, and Finland, whose democratic credentials and consensus-orientation in politics in the post-war period no one will dispute, have all successfully used industrial policy in one way or another. In fact, one can go one step further to argue that, if there is a democratic consensus on it (as in the above-mentioned countries), industrial policy can be even more effectively implemented. After all, the long-term success of every public policy requires some degree of consent by those affected by it (see Weiss, 1998, ch. 2, for a similar argument).

The fact that industrial policy in Taiwan, and especially Korea, has lost its political legitimacy in the eyes of the population because of its past association with dictatorship has created the impression that democracy and industrial policy are mutually incompatible. However, there is no inherent reason why industrial policy cannot regain legitimacy even in these countries, if a democratic political consensus can be built around it. While it may be argued that there is no chance of that happening in the near future, at least in Korea, this is an argument based on an assessment of the present political situation in the country. This should be distinguished from the argument that industrial policy is inherently incompatible with democracy.

To summarise, the association between industrial policy and authoritarianism in the minds of some observers of certain East Asian countries is understandable, but this association is due to a specific historical conjuncture, rather than something inevitable. If we applied the same logic to the experiences of nineteenth-century Britain, USA, or some other European countries, we would have probably concluded that a free market was incompatible with democracy. This was indeed what most liberals of the time thought to be the case (on the earlier view on the relationship between democracy and liberalism, see Bobbio, 1990; see also Polanyi, 1957).

Changing government–private sector balance of power

Throughout their economic development, but especially more recently, many East Asian countries have witnessed the rise of large private sector industrial and financial corporations, and their increasing internationalisation. This has prompted many people to argue that industrial policy that restricts private sector interests will not be possible any more, as private sector firms now have enormous political influence. Besides their weight in the economy, they also have the ability to veto government policy by 'exiting' from the domestic economy.

This argument obviously makes a lot of sense. Corporations which have become economically and politically more powerful and have more freedom to move around the world certainly would be, other things being equal, better able to resist government policies that sacrifice their individual interests for the sake of national goals. And indeed, in Korea, the giant conglomerates (*chaebols*) aggressively campaigned during the 1990s to convince the population that the government should abandon industrial policy and financial regulations.[15] However, one needs to be careful in jumping from such observations to the conclusion that economic development means the rise of the private sector, which, when combined with globalisation, makes industrial policy impossible.[16]

There is no clear and inevitable association between economic development, the rise of the private sector, and the demise of industrial policy. The experience of Taiwan shows that economic development need not lead to the emergence of a powerful private sector, as the process of corporate development is as much a political process as an economic process (Fields, 1995). The Japanese experience is also consistent with such an observation. The Japanese corporations had already become very powerful and internationally mobile during the 1970s and the 1980s, but Japan had great success with industrial policy during that period, because these firms accepted the legitimacy of industrial policy and co-operated with the government to ensure its success.

Moreover, it needs to be pointed out that the extent of internationalisation of even the largest corporations in East Asia is still limited. From the experiences of other countries with longer histories of internationalised business, the chances that they will turn into truly 'transnational' corporations without a 'home base' in the foreseeable future is low. When we also note that globalisation is a trend that can be, and has been (in the aftermath of the Great Depression) reversed, it is not clear whether the current process of globalisation will continue until it makes industrial policy impossible.[17]

So, in the end, the argument that industrial policy has become less feasible in East Asian countries because economic development has led to the growth in power of the private sector, which naturally resists industrial policy, is problematic. It may fit the Korean example rather well, but as a general proposition it is rather suspect. This is because there is no direct causal link between economic development and the rise of the private sector (recall the Taiwanese counter-example), on the one hand, and between the rise of the private sector (including its internationalisation) and the demise of industrial policy (recall the Japanese counter-example), on the other.

Concluding remarks

So what are the main conclusions we can draw from 're-thinking' the issue of industrial policy in East Asia?

First of all, there are more theoretical justifications for industrial policy than acknowledged by the *EAM*, which need to be explored deeper. This is important, especially given that these justifications (e.g. co-ordination of competing investments, scale economies) were probably more important in the actual formation of industrial policy in the East Asian countries than the ones acknowledged by the *EAM* (e.g. 'Big Push', formation of implicit cartels in international negotiations).

Second, better ways to test the true effects of industrial policy are needed. Apart from the detailed methodological criticisms that have already been made of the tests conducted on the issue reported in the *EAM*, there remain some issues that need to be resolved in future research. For example, how should we deal with the problem of externalities? Also, whether and how do we quantify the effects of policies such as the achievement of scale economies through licensing policy, the prevention of a price war through encouragement of a recession cartel, or the reduction in resistance to technological change in the long run through the use of 'protective' industrial policy? These are only some of the questions to be explored further.

Third, the capability argument used by the *EAM* against other countries wanting to adopt East-Asian-style industrial policy is not without its merits, but it has important limitations. First, it is not clear why industrial policy, regardless of its scale and sophistication, requires an exceptionally competent bureaucracy. And, second, bureaucratic capability is something that can be developed through deliberate efforts and learning-by-doing. It should also be added that it is not as if more market-oriented systems do not require high institutional capabilities, as we can see in the difficulties that many transitional and developing economies are currently experiencing in establishing a 'free-market system'.

Fourth, the WTO argument against the feasibility of industrial policy in the present era, also emphasised by the *EAM*, draws an overly pessimistic conclusion without looking at the full array of possibilities that exist for policy manoeuvre.

Fifth, as for the argument that the recent recession in Japan and crises in other East Asian countries prove that industrial policy has ultimately been detrimental for these economies, it should be pointed out that the countries which did not have or had ditched industrial policy were actually in crisis.

Finally, as for things like economic maturity, democratisation and the rise of private sector power, frequently cited as reasons why industrial policy will not be feasible any more, even in countries that have successfully used it, these presuppose causal links that are neither robust nor really backed by empirical evidence.

Acknowledgements

I thank Farrukh Iqbal for his helpful suggestions in the design of the chapter. Han-Kyun Rho provided valuable research assistance. I thank the following participants at the World Bank workshop where an earlier version of this chapter was presented for their helpful comments: Migara De Silva, Howard Pack, Stephen Parker, Dwight Perkins, Meredith Woo-Cumings, and especially Joseph Stiglitz. I also benefited from comments by Stanley Engerman, Ben Fine, Jomo K.S., Kyun Kim, Jonathan Pincus, Alfredo Saad Filho and Ajit Singh.

Notes

1 Important works that emphasised the positive contribution of Japanese industrial policy include Magaziner and Hout (1980), Johnson (1982), and Reich (1982). In the opposite camp, Schultze (1983) and Badaracco and Yoffie (1983) were influential. Reviews of this first phase of the industrial policy debate can be found in Johnson (1984), Thompson (1989), and Chang (1994, ch. 3).

2 For a more detailed review of the evolution of the debate on industrial policy leading up to the publication of the *East Asian Miracle* report, see Chang (1993). For the political background to the publication of the report, see Wade (1996).

3 For example, as late as 1988, the famous mainstream trade economist Bela Balassa was arguing that 'apart from the promotion of shipbuilding and steel, [the role of the state in Korea] has been to create a modern infrastructure, to provide a stable incentive system, and to ensure that government bureaucracy will help rather than hinder exports' (Balassa, 1988, S.286).

4 Of course, this does not mean that new entries cannot or do not happen. East Asian producers have been quite good at gaining market shares in some industries with chronic over-capacity problems. However, successful entry into these industries will be much more difficult than entry into other industries.

5 When redundant investments emerged for whatever reason (e.g. government failure to take timely action, non-compliance by firms), the East Asian governments tried to minimise excess capacity by encouraging, and sometimes forcing, mergers or recession cartels. There have, of course, been national differences. The Japanese government

preferred to use recession cartels, while the Korean government periodically resorted to forced mergers. Examples of the latter include the so-called 'industrial restructuring programme' of the early 1980s (which affected industries such as automobile, naval diesel engines, copper smelting, power-generating equipment, heavy electrical machinery, and electronic switching systems) and the so-called 'Big Deal' programme following the 1997 crisis (which affected industries such as semiconductors, automobiles, power-generating equipment, naval diesel engines, aircraft, petrochemical, petroleum refining, and railway carriages).

6 The cost inefficiency that results from the presence of excessive product variety is widely recognised in the car industry. For example, in South Africa and Taiwan, about ten manufacturers each produce several thousand cars annually in an industry where the minimum efficient scale is believed to be around 300,000 units per year.

7 I thank Joe Stiglitz for highlighting this dimension of protective industrial policy.

8 Different countries have dealt with this problem in different ways. Many European economies have used unemployment insurance to soften the blow of structural change on the owners of specific human skills and *ad hoc* subsidies to do the same for the owners of specific physical equipment. More proactively, the Scandinavian countries combined such systems with 'active labour market policy', which provided re-training and relocation subsidies to workers. East Asian countries used protective industrial policy to deal with this problem.

9 The *EAM* distinguishes industrial policy as a separate category from export promotion policy. However, this is not right because export promotion was a key element in the industrial policy regimes of East Asian countries. New industries that the governments wanted to promote almost invariably needed access to foreign exchange to buy new technologies and equipment embodied by them; knowing this, the government saw export success as a prerequisite for industrial upgrading. See also Rodrik (1994).

10 Paying attention to these hitherto-ignored aspects of industrial policy makes empirical testing more difficult. Traditionally, many studies have used indicators such as subsidies and tariffs to measure industrial policy in an industry. However, recognising these additional aspects also means taking into account less quantifiable factors like costs saved due to co-ordination of competing investments and from measures to achieve scale economies. Gains from the acceleration of structural change that protective industrial policy may produce will be even more difficult to measure, as these are likely to spill over into the rest of the economy.

11 The *EAM* also cites 'the pragmatism and flexibility of governments' as another condition, but this is actually less important in the argument.

12 The list of 'fundamentals' in the *EAM* keeps changing, because it does not really have a good theory of which policy area is more important and why. However, these three items are almost always included in the list.

13 For example, the *EAM* does recognise that there is a time provision of up to eight years for developing countries to bring their trade policies in line with those practised in advanced countries (p. 365). It also acknowledges that there are means other than subsidies or export-directed credit programmes that may be used to promote exports (p. 25).

14 For some comprehensive discussions of the East Asian crisis, see Furman and Stiglitz (1998), Radelet and Sachs (1998), Stiglitz (1998), Singh (1999), and Chang (2000).

15 In 1996, the Korean Federation of Industries, the association of *chaebols*, prepared a report arguing for the abolition of all government ministries except the ministries of defence and foreign affairs and for the consequent reduction of government staff by 90 per cent. The report had to be officially withdrawn because it was leaked in advance by a careless reporter and created a popular uproar. While the chance of such a proposal being taken seriously was non-existent even in Korea, then (and still to a large extent) in the grip of an anti-statist reaction, the incident is illustrative of the

aggressiveness of the *chaebols* in pushing for greater business freedom in the recent period.

16 For a comprehensive critique of the argument that globalisation makes industrial policy impossible, see Weiss (1998). Chang (1998a) makes a similar argument more specifically in relation to the rise of transnational corporations.

17 In the late nineteenth and the early twentieth centuries, the world economy was almost as globalised as it is now on many indicators, and even more on some indicators. For example, international labour mobility was much higher and international policy uniformity was much greater then – especially given the Gold Standard and the lack of tariff autonomy in all countries except the strongest (even Japan did not have tariff autonomy until 1911). See Bairoch and Kozul-Wright (1996) for historical evidence.

References

Akyuz, Y., H.-J. Chang and R. Kozul-Wright (1998) 'New perspectives on East Asian development', *Journal of Development Studies*, 34(6).

Amsden, A. (1985) 'The state and Taiwan's economic development', in P. Evans, D. Rueschemeyer and T. Skocpol (eds) *Bringing the State Back In*, Cambridge: Cambridge University Press.

—— (1989) *Asia's Next Giant*, New York: Oxford University Press.

Amsden, A. and A. Singh (1994) 'The optimal degree of competition and dynamic efficiency in Japan and Korea', *European Economic Review*, 38(3/4).

Badaracco, J. and D. Yoffie (1983) 'Industrial policy: it can't happen here', *Harvard Business Review*, November/December.

Bairoch, P. and R. Kozul-Wright (1996) 'Globalisation myths: some historical reflections on integration, industrialisation and growth in the world economy', UNCTAD Discussion Paper, no. 113, Geneva: United Nations Conference on Trade and Development (UNCTAD).

Balassa, B. (1982) 'Development strategies and economic performance', in B. Balassa *et al.*, *Development Strategies in Semi-Industrial Economies*, Baltimore, MD: Johns Hopkins University Press.

—— (1988) 'The lessons of East Asian development: an overview', *Economic Development and Cultural Change*, 36(3), April, Supplement.

Bhagwati, J. (1988) *Protectionism*, Cambridge, MA: MIT Press.

Bobbio, N. (1990) *Liberalism and Democracy*, London: Verso.

Chang, H.-J. (1993) 'The political economy of industrial policy in Korea', *Cambridge Journal of Economics*, 17(2).

—— (1994) *The Political Economy of Industrial Policy*, London and Basingstoke: Macmillan.

—— (1995) 'Explaining "Flexible Rigidities" in East Asia', in T. Killick (ed.) *The Flexible Economy*, London: Routledge.

—— (1997a) 'Luxury consumption control and industrialisation in East Asia', mimeo, a background paper prepared for *Trade and Development Report 1997*, Geneva: UNCTAD.

—— (1997b) 'The economics and politics of regulation', *Cambridge Journal of Economics*, 21(6).

—— (1998a) 'Transnational corporations and strategic industrial policy', in R. Kozul-Wright and R.E. Rowthorn (eds) *Transnational Corporations and the World Economy*, London and Basingstoke: Macmillan.

—— (1998b) 'Korea: The misunderstood crisis', *World Development*, 26(8).

—— (2000) 'The hazard of moral hazard – untangling the Asian crisis', *World Development*, 28(4).

Chang, H-J. and R. Rowthorn (1995) 'Role of the state in economic change: entrepreneurship and conflict management', in H.-J. Chang and R. Rowthorn (eds) *Role of the State in Economic Change*, Oxford: Oxford University Press.

Chang, H.-J., H.-J. Park and C.G. Yoo (1998) 'Interpreting the Korean crisis: financial liberalisation, industrial policy, and corporate governance', *Cambridge Journal of Economics*, 22(6).

Chang, H.-J. and P. Evans (2000) 'The role of institutions in economic change', paper presented at a workshop of 'The Other Canon' Group, 13–14 January, Venice, Italy.

Cheng, T., S. Haggard and D. Kang (1998) 'Institutions, economic policy, and growth in the Republic of Korea and Taiwan Province of China', *Journal of Development Studies*, 34(6).

Dore, R. (1986) *Flexible Rigidities: Industrial Policy and Structural Adjustment in the Japanese Economy, 1970–80*, London: The Athlone Press.

Evans, D. and P. Alizadeh (1984) 'Trade, industrialisation, and the visible hand', *Journal of Development Studies*, 21(1).

Evans, P. (1995) *Embedded Autonomy – States and Industrial Transformation*, Princeton: Princeton University Press.

Fields, K. (1995) *Enterprise and State in Taiwan and Korea*, Ithaca, NY: Cornell University Press.

Financial Times (2000) *Markets 2000*, supplement, January.

Fishlow, A., C. Gwin, S. Haggard, D. Rodrik and R. Wade (1994) *Miracle or Design? – Lessons from the East Asian Experience*, Washington, DC: Overseas Development Council.

Fransman, M. (1990) *The Market and Beyond: Information Technology in Japan*, Cambridge: Cambridge University Press.

Furman, J. and J. Stiglitz (1998) 'Economic crises: evidence and insights from East Asia', Brookings Papers on Economic Activity, no. 2.

Henderson, J. (1999) 'Uneven crises: institutional foundations of East Asian economic turmoil', *Economy and Society*, 28(3).

Johnson, C. (1982) *MITI and the Japanese Miracle*, Stanford: Stanford University Press.

—— (ed.) (1984) *The Industrial Policy Debate*, San Francisco: Institute for Contemporary Studies.

Jomo, K.S. and M. Rock (1998) 'Economic diversification and primary commodity processing in the second-tier South-East Asian newly industrialising countries', UNCTAD Discussion Paper, no. 136, Geneva: United Nations Conference on Trade and Development (UNCTAD).

Jones, L. and I. Sakong (1980) *Government, Business and Entrepreneurship in Economic Development: The Korean Case*, Cambridge, MA: Harvard University Press.

Kwon, J. (1994) 'The East Asia challenge to neoclassical orthodoxy', *World Development*, 22(4).

Lall, S. (1994) 'Does the bell toll for industrial policy?', *World Development*, 22(4).

Little, I. (1982) *Economic Development*, New York: Basic Books.

Luedde-Neurath, R. (1986) *Import Controls and Export-Oriented Development: A Reassessment of the South Korean Case*, Boulder, CO, and London: Westview Press.

Magaziner, I. and T. Hout (1980) *Japanese Industrial Policy*, London: Policy Studies Institute.

Okimoto, D. (1989) *Between MITI and the Market: Japanese Industrial Policy for High Technology*, Stanford: Stanford University Press.

Polanyi, K. (1957) *The Great Transformation*, Boston, MA: Beacon Press.

Radelet, S. and J. Sachs (1998) 'The East Asian financial crisis: diagnosis, remedies and prospects', *Brookings Papers on Economic Activity*, no. 1.

Ranis, G. and J. Fei (1975) 'A model of growth and employment in the open dualistic economy: the cases of Korea and Taiwan', in F. Stewart (ed.) *Employment, Income Distribution, and Development*, London: Frank Cass.

Reich, R. (1982) 'Why the US needs an industrial policy', *Harvard Business Review*, January/February.

Renshaw, G. (1986) *Adjustment and Economic Performance in Industrialised Countries: A Synthesis*, Geneva: International Labour Office (ILO).

Rodrik, D. (1994) 'King Kong meets Godzilla', in A. Fishlow *et al.*, *Miracle or Design? – Lessons from the East Asian Experience*, Washington, DC: Overseas Development Council.

Schultze, C. (1983) 'Industrial Policy: A Dissent', *The Brookings Review*, Fall.

Singh, A. (1994) '"Openness" and the "market-friendly" approach to development: learning the right lessons from development experience', *World Development*, 22(12).

—— (1999) 'Asian capitalism and the financial crisis' in J. Michie and J. Grieve Smith (eds) *Global Instability – The Political Economy of World Economic Governance*, London: Routledge.

Stiglitz, J. (1998) 'Sound finance and sustainable development in Asia', keynote address to the Asia Development Forum, jointly organised by the World Bank and the Asian Development Bank, 9–12 March, Manila, the Philippines.

Telser, L. (1987) *A Theory of Efficient Cooperation and Competition*, Cambridge: Cambridge University Press.

Thompson, G. (ed.) (1989) *Industrial Policy: US and UK Debates*, London: Routledge.

Wade, R. (1990) *Governing the Market*, Princeton: Princeton University Press.

—— (1996) 'Japan, the World Bank, and the art of paradigm maintenance: the East Asian Miracle in political perspective', *New Left Review*, 217, May/June: 3–36.

Weiss, L. (1998) *The Myth of the Powerless State*, Cambridge: Polity Press.

Westney, E. (1987) *Imitation and Innovation: The Transfer of Western Organisational Patterns to Meiji Japan*, Cambridge: Cambridge University Press.

Williamson, O. (1985) *The Economic Institutions of Capitalism*, New York: The Free Press.

World Bank (1987) *World Development Report 1987*, New York: Oxford University Press.

—— (1993) *The East Asian Miracle: Economic Growth and Public Policy*, New York: Oxford University Press.

—— (1997) *World Development Report 1997*, New York: Oxford University Press.

—— (1999) *World Development Report 1999–2000*, New York: Oxford University Press.

3 Behind POSCO'S success

The role of government in technology capability building

Park Eul Yong

How do we explain the rapid and sustained growth of Asia's newly industrialising economies (NIEs) since the 1960s? During the past three decades, Korea, Taiwan, Singapore and Hong Kong grew at an average of 8–9 per cent annually in real terms. Such growth was unprecedented in modern economic history. What factors made the rapid and sustained growth of the NIEs possible? The spread of rapid and sustained economic growth to Southeast Asian countries and to China in the 1980s, and to Vietnam in recent years, makes the task of probing the factors responsible for the sustained long-term growth of East Asian economies even more challenging.

There are numerous studies on this subject,[1] and a number of important factors have been identified as causes of the rapid and sustained growth of the NIEs. First, a rapid growth of capital stock; second, large investments in human capital; and third, emphasis on linkages with foreign markets through exports, and with foreign technology and capital. Despite the substantial research already conducted, however, there are two important issues on which economists sharply disagree, and which, therefore, still need to be probed further. The first of these is the role of government. It is fairly well recognised that government played an active role in three of the four Asian NIEs, namely in Korea, Taiwan and Singapore, with Hong Kong being the exception. The question is: did these countries grow quickly for a long period of time because of, or in spite of, the active role of their governments? The second issue is whether the rapid and sustained growth of the Asian NIEs can be accounted for by the supply of factors, especially capital and labour, or by the growth of productivity through innovation. Recently, Krugman (1994) and Young (1995) argued that the rapid growth of Asian NIEs can be largely explained by factor supply, rather than productivity growth through innovation, as can be seen in the experience of industrial countries.

The purpose of this chapter is to shed light on these two challenging issues through a case study of POSCO, a large integrated steel company in Korea. In addition, we would like to discuss whether the experience of the Asian NIEs can become a model for other developing countries to emulate in order to achieve sustained growth. If this is not the case, we need to know why the Asian NIEs' experiences may not be relevant for other developing countries.

The organisation of this chapter is as follows. Following the introduction, two key issues related to market failure and the role of government, and to innovation, learning and productivity growth will be discussed. In the third part, the development history of POSCO and changes in its performance and competitiveness are discussed, along with the rapid growth of the company's productivity. Part four studies the role of learning and technology development at POSCO to explain its excellent performance and competitiveness. The fifth part discusses the role of government in the development of POSCO, focusing in particular on making POSCO an internationally competitive firm in steel making. Finally, we argue that a social capability approach, which goes beyond factor supply and the availability of institutions and technology, is necessary to understand the rapid and sustained growth of the NIEs. We will thus be able to understand why the right approach for Korea and other Asian NIEs to achieve sustained long-term growth may not be applicable to other developing countries.

Role of government in the economic development of the NIEs

The role of government in the rapid and sustained growth of East Asian NIEs has been one of the most contentious issues among economists. Neo-classical economists maintain that the phenomenal growth of the NIEs can largely be explained by the active role of the private sector and well-functioning market systems. According to them, stable macroeconomic conditions, high rates of savings and investment, investment in education and human capital, and liberalisation of foreign trade and investment were important in promoting development through active markets. Although some neo-classical economists admit the necessity of a relatively active government role in certain areas, such as export promotion and facilitation of technological capability development, they interpret this as market-friendly government action.[2]

The alternative 'development state' view maintains that the role of government has been an essential ingredient of the NIEs' development strategy (see Amsden, 1989; Wade, 1990). According to this perspective, without the role of government to induce entrepreneurs to invest in high-risk ventures as well as to compete in international markets with incentives, subsidies, protection and contingent rents (i.e. rents for entrepreneurs conditional on desired performance), the rapid and sustained growth of the NIEs might not have been realised.

The conventional neo-classical economic view holds that government intervention in the market should be limited to the correction of market failures, e.g. in cases of monopoly, externalities, public goods and asymmetries of information. Neo-classical economists maintain that the proper role of government includes: maintaining macroeconomic stability, promoting growth and employment, providing infrastructure and public goods (such as education, national security, legal system and basic research), and so on.

Government intervention in the market to correct market failures usually assumes the existence of well-functioning markets. However, in many developing countries, markets have not been well developed. In Asia, governments have supported and helped the private sector and the market system to develop. Thus, an entirely different role for government emerged in the Asian developing countries, namely for the promotion of markets and private sector development. In latecomer countries like Japan, Korea and Taiwan, the government initially played a very important role in promoting the development of markets and the private sector since markets in these countries were weak or poorly developed.

Those who argue that the active role of government has been the critical factor in the rapid and sustained growth of the NIEs do not deny that the private sector and market system have been essential, but rather that the government has played a significant role, especially when markets were weak and not well developed. Those who hold the view that the market system, and not the government, was primarily responsible for the rapid growth of the NIEs admit that governments sometimes played an important role in such countries as Korea, Taiwan, Singapore and Japan during their initial years of development, even though government interventions in the market system also created distortions and inefficiencies. Therefore, it is not useful to argue over whether or not the role of government was very important for the sustained and rapid economic development of late developers. The debate must go beyond that. We need to study when, not whether, and under what conditions government intervention in the market system worked, and when it did not work.

Recently, a new view has emerged concerning the role of government in developing countries, namely the market enhancing role of the government (Aoki *et al.*, 1995). This view is more appealing than others because it takes closer cognisance of the facts involved. The market-enhancing role of government is 'to facilitate the development of private sector institutions that can overcome market failures' (Aoki *et al.*, 1995). The difference between the 'market enhancing' view on the one hand and the 'market friendly' and 'developmental state' views are as follows. The market-friendly view looks to private sector institutions to resolve most market failures and imperfections, while the developmental state view sees government intervention as the principal solution to such problems. The new view, however, looks at government as helping developing markets and private sectors to correct market failures in developing countries.

The case of POSCO is interesting, first of all, because the underdevelopment of the financial market in Korea during the early period of development led the government to play an active role in high-risk ventures such as POSCO. Second, although POSCO started as a state-owned enterprise, its management and operations were independent from government intervention, and the management could run the company as a private firm. By closely examining the case of POSCO, we can analyse the role of government to determine when and why an active role of government worked.

Role of innovation and learning

In recent years, a number of studies (Krugman, 1994; Young, 1995) have argued that the rapid growth of East Asian NIEs can largely be explained by the rapid growth of key factor inputs such as physical and human capital. According to these studies, the role of innovation to promote productivity growth did not play a significant role in the rapid growth of East Asian NIEs. The implication of these studies is that unless NIEs first invest heavily in building technological capability and promoting productivity growth through active innovation, the growth of Asian NIEs may slow down in the near future, as the capability to invest continually in physical capital stock and human capital at such a rate may not be possible.

Nelson and Pack (1995), however, argued that assimilation through innovation and learning as well as accumulation of key factors have played important roles in the rapid growth of the NIEs. They also pointed out various problems in the methodology of the growth accounting approach that the studies by Krugman (1994) and Young (1995) used. The main issues are whether rapid growth in the investment of physical and human capital could have been possible without innovation and learning in the NIEs, and how to interpret the technology development in the NIEs in terms of innovation. Let us therefore review recent literature on innovation and learning in the context of developing countries to understand the nature of technology development in developing countries.

Technology development in developing countries

Developing countries have at least one advantage as follower countries in pursuit of industrialisation. They can utilise technologies and know-how already available in industrial countries rather than trying to 're-invent' them. This provides significant advantages for developing countries as the time and cost required to absorb technology and know-how already available in industrial countries are generally only a fraction of what is required to 're-invent' them. However, many case studies have shown that it is not possible for a developing country to absorb foreign technology without strong absorptive capability and a concerted development effort.

Numerous studies have identified the following three steps in the efficient utilisation of foreign technology and know-how by a developing country (Westphal *et al.*, 1985; Kim, 1977: 209–11). The first step is for users in developing countries to learn by copying and imitating technology from industrial countries. It is important to note that even copying and imitating are not necessarily easy for a latecomer. Technology transfer is usually limited to formal, codified and embodied technology without the tacit elements often involved. The sellers of technology do not usually provide assistance to help buyers learn all the small but crucial steps involved in particular product or process technologies. The buyers of technology must have the technological capability to acquire, adapt and improve imported technology. It can be argued, therefore, that there is no exact copying of a product or a process technology in reality. The process of copying and imitating usually requires serious efforts of development and absorption on the part of imitators. Frequently, this step also involves an understanding of basic technological principles which may serve as the basis for the adaptation and improvement of imported technology.

The second step in the development of technology in developing countries involves adaptations and small improvements to imported technology. The degree of success in such efforts depends on the importers' technological capability for adaptation and improvement. In this stage, significant differences in the pattern of technological development among different developing countries emerge. Some countries, such as Korea and Taiwan, have made explicit policy efforts to learn actively through acquisition, adaptation and internalisation of imported technology. Through various policy measures, governments promote the development of technology in close co-operation with the private sector. Active efforts to absorb foreign technology have been a critical part of their own programmes to develop their own technology and know-how. Other countries have made only passive efforts at catching-up by relegating the main responsibility to multinational firms or by adopting *laissez-faire* policies.

The third and final step in absorbing imported technology involves innovation through internalisation. Innovations take place more readily through the internalisation of imported technology and know-how. Furthermore, learning through imitating, adapting, improving and developing leads to innovation in developing countries. In the real world, clear distinctions among different successive steps in the technological development process may not exist, i.e. learning and innovation may take place at the same time.

Learning and innovation in developing countries

As noted, in developing countries, technology generally changes through the process of technological learning; i.e. through acquisition, adaptation and improvement of existing technologies in industrial countries, rather than through innovation at the frontiers of science and technology. It does not necessarily mean that innovations do not take place in the context of developing countries. In the traditional literature, learning and innovation were understood as completely different concepts. More careful case studies of industrial development in developing countries in recent years reveal that activities such as imitation, copying, adaptation and improvement entail small innovations and development on the part of firms that acquire technology. From this point of view, innovation and learning are very closely related. Small innovations constantly take place in economic activities as a result of learning, such as learning by manufacturing, learning by using, and interaction among various agents in the market (producers and users, for example). This view is particularly important for developing countries because the conventional view says that the industrialisation of developing countries is supposed to be based on learning existing technologies of industrial countries, whereas that of industrial countries is ostensibly based on 'innovation'. Thus, the fact that many small innovations occur during the process of learning and absorbing imported technology has not been properly understood. Many studies of industrial technology development in Japan and the NIEs suggest that the accumulation of such small innovations have been very critical for Japan's technological leadership in recent years and in the sustained rapid growth of the NIEs (Kim, 1977; Lall, 2000; Hobday, 2000).[3]

When the incremental and cumulative aspects of innovation are recognised, the role of organisational learning becomes critical.[4] A firm strong in organisational learning capability will be more efficient in the incremental and cumulative aspects of innovation. The organisational learning set-up helps to distribute knowledge within the firm and to promote organisation-wide innovation.

A number of economic conditions affect the learning process in developing countries, such as the size and growth of demand for the products for which the learning takes place. Increasing demand for such products provides the opportunity to accelerate 'learning by manufacturing' as well as cumulative improvements in productivity growth through imitation, adaptation and improvement in technology. Competition is another important condition that affects the learning process.[5] This is the reason why exports are important for effective and efficient learning, since it increases demand and requires the firm to face competition. Another important condition which affects the learning process is government policy. Government policy affects learning and innovation through trade policy, R&D policy as well as macroeconomic policy.[6]

POSCO: development and performance

The development experience of the Pohang Iron and Steel Company (POSCO) is remarkable.[7] Despite Korea's initial lack of technology, experience and capital to build and operate an integrated steel mill, POSCO became the second largest steel mill and one of the most efficient steel producers in the world within two decades after its start-up in 1973.

The promotion of an integrated steel mill in Korea in the 1960s was a difficult task as Korea lacked nearly all the necessary key elements to build and operate a modern integrated steel plant. Korea was a poor country whose per capita income was around US$100 in the mid-1960s. The small size of the economy as well as its poor economic growth track record did not convince financiers and specialists that there was sufficient steel demand in Korea to build an integrated steel plant with a minimum efficient scale of one million tons capacity. Moreover, Korea did not have sufficient capital to finance this capital intensive project, nor did it have the technology and know-how to build and operate an integrated steel plant. Even though the necessary process technology could be imported from the suppliers of steel-making machinery and equipment, successful operation of an integrated steel mill requires engineering skills that Korea did not have at the time. Also, Korea had not produced any good quality iron ore or coke, so all raw materials had to be imported from abroad. For these reasons, there were numerous objections by professional experts from abroad to the Korean government's plan to build an integrated steel mill.[8] Efforts to finance the ambitious new project failed a few times in the early 1960s due to disagreements between foreign financiers and the Korean government concerning the scale of the steel mill and other issues. Opposition by international experts was understandable considering the unsuccessful attempts to build efficient integrated steel mills in India, Mexico, Brazil and Turkey.

Despite these objections and seemingly unfavourable conditions for building an efficient integrated steel mill, the Korean government proceeded with its plan. Korea's Second Five-Year Economic Development Plan (1967–71) listed construction of an integrated steel mill and petrochemical plants as its major industrial development projects. In 1967, the government passed a new law to promote an integrated steel mill, the Law for the Promotion of the Steel Industry. Finally, POSCO was established in 1968, with the government-appointed retired general T.J. Park, who had successfully run a large state-owned enterprise, Korea Tungsten Company, as president.

In 1969, the Korean government was able to persuade Japan to provide the necessary technology and a commercial loan (US$67 million) to purchase Japanese machinery and equipment. In addition, the Korean government decided to use 25 per cent of the Japanese reparations fund (US$77.2 million) for the steel project. Austria also provided a commercial loan to sell its machinery and equipment (US$30 million). The remaining capital was mobilised through share subscriptions by the Ministry of Finance, the Korea Development Bank and the Korea Tungsten Company, and with loans from Korean banks. Eventually, the entire POSCO investment in the Pohang plant reached 5,554 billion won. Later, POSCO built new steel production facilities in Kwangyang during 1985–92, investing 8,459 billion won there,[9] the largest ever investment in Korean industrial history.

To build and operate the integrated steel mill, the Korean government decided to establish a new state-owned enterprise (SOE), the Pohang Iron and Steel Company (POSCO). As an SOE, POSCO has been an exception in various regards. Nearly all other major import substitution activities supported by the government, such as shipbuilding, refining, automobile and machine manufacture, were allocated to private sector firms, usually one of the *chaebols*. There were several major reasons for such decisions. First, the required investment was too large and hence too risky for a private sector firm to invest, even with substantial government support. Second, steel is an important intermediate good required by other industries such as shipbuilding, automobile and machinery, and POSCO was likely to be a monopoly for some time to come due to scale economy considerations and the limited domestic demand. Therefore, allocating the integrated steel mill to a private sector firm could cause problems. Third, the government had to consider the substantial size of the government support required for the project in the initial period, and decided that it would be better to provide this to a state-owned enterprise, rather than a private enterprise. Despite the decision to make POSCO an SOE, it was different from other SOEs in Korea. POSCO was set up under the Commercial Law of Korea as a joint stock company, with its board of directors and stockholders, not the government, running the company.[10] The independence of POSCO's management from government intervention was assumed in advance.

Table 3.1 Capacity additions and plant specifications of POSCO, 1970–92

			Phases at Pohang		
			I	*II*	*III*
Period of construction			Apr. 1970–July 1973	Dec. 1973–May 1976	Aug. 1976–Dec. 1978
Capacity (1,000 ton/year)			1,030	2,600	5,500
S	Sintering plant	Spec.	DL type 130m	DL type 204m	DL type 400m
P		Cap.	1,322,000 T/Y	2,197,000 T/Y	4,292,000 T/Y
E	Coke oven	Spec.	68 ovens	106 ovens	146 ovens
C		Cap.	584,000 T/Y	912,000 T/Y	1,552,000 T/Y
I	Blast furnace	Spec.	1,660m	2,550m	3,795m
F		Cap.	1,011,000 T/Y	1,697,000 T/Y	2,752,000 T/Y
I	Steel making	Spec.	100 tons/heat	100 tons/heat	300 tons/heat
C			X 2	X 1	X 2
A		Cap.	1,032,000 T/Y	(2,000,000 T/Y)	3,300,000 T/Y
T I	Continuous casting	Spec.	—	4 strand X 1 machine	—
O		Cap.	—	1,026,000 T/Y	—
N	Hot-strip mill	Spec.	RF 150 ton/hour	RF 150 ton/hour	RF 150 ton/hour
			X 1	X 1	X 1
O		Cap.	606,500 T/Y	(775,500 T/Y)	(1,410,000 T/Y)
F	Cold-strip mill	Spec.	—	TCM, CGL	—
		Cap.	—	71,100 T/Y	—
P	Plate mill	Spec.	RF 100 ton/hour	—	RF 235 ton/hour
L			X 1		X 1
A N		Cap.	336,000 T/Y	—	1,243,000 T/Y
T	Wire rod mill	Spec.	—	—	2 strand
S	Cap.	—	—	446,000 T/Y	—

Source: Amsden (1989: 294, Table 12.1).
Notes: Spec.: specification; Cap.: capacity; T/Y: ton/year;
( ) = capacity after expansion.

		Phases at Kwangyang			
IV–1	*IV–2*	*I*	*II*	*III*	*IV*
Feb. 1979– Feb. 1981	Sept. 1981– May 1983	March 1985– June 1987	Nov. 1986– Dec. 1988	Nov. 1988– Dec. 1990	Jan. 1991– Oct. 1992
8,500	9,100	2,700	5,400	8,100	11,400
DL type 400m		DL type 400m	DL type 400m	DL type 400m	DL type 424m
4,292,000 T/Y		4,426,000 T/Y	4,426,000 T/Y	4,426,000 T/Y	4,650,000 T/Y
150 ovens	75 ovens	132 ovens	132 ovens	132 ovens	132 ovens
1,552,000 T/Y	733,000 T/Y	1,430,000 T/Y	1,430,000 T/Y	1,430,000 T/Y	1,430,000 T/Y
3,759m	II Relining	3,800m	3,800m	3,800m	3,800m
2752000 T/Y	—	2,840,000 T/Y	2,840,000 T/Y	2,840,000 T/Y	2,910,000 T/Y
300 tons/heat	—	250 tons/heat	250 tons/heat	250 tons/heat	250 tons/heat
X 1		X 2	X 1	X 2	X 1
(6,500,000 T/Y)	—	2,784,000 T/Y	(5,568,000 T/Y)	2,784,000 T/Y	
2 strand X 2 machine	—	2 strand X 2 machine	2 strand X 1 machine	2 strand X 2 machine	2 strand X 2 machine
(3,844,000 T/Y)	—	2,700,000 T/Y	(5,400,000 T/Y)	3000000	(6,000,000 T/Y)
RF 250 ton/hour	—	RF 300 ton/hour	RF 300 ton/hour	RF 270 ton/hour	RF 230 ton/hour
X3		X 2	X 1	X 2	X 3
3,311,000 T/Y	—	2,660,000 T/Y	(4,433,000 T/Y)	2,967,000 T/Y	3,500,000 T/Y
TCM, CAL	—	—	ROL, PCM, CGL	PCM, PGL, CAL	PCM, CAL, No CGL
1,000,000 T/Y	—	—	1,675,000 T/Y	1,622,000 T/Y	1,225,000 T/Y
—	—	—		—	
—	—	—	—	—	—
—	1 strand	—	—	—	—
350,000 T/Y	—	—	—	—	

The first phase began in 1970 and construction took only 27 months to complete. The mill, with a capacity of one million tons, was to consist of a sintering plant, coke oven, blast furnace, basic oxygen furnace and hot-rolled mill. Immediately after the completion of the first phase, second phase construction, with an additional capacity of 1.6 million tons, started in December 1973. This added a continuous casting and cold-strip mill to the same plant capacities developed in Phase I. The construction of Phase II only five months after completion of Phase I reflected strong confidence in efficiently building and operating the integrated steel mill to compete with foreign steel producers. By 1983, POSCO continually added new capacity through the second stage of Phase IV to reach a total capacity of 9.1 million tons per year. The successful operation of POSCO and rapidly increasing domestic demand for steel convinced POSCO and the government to further expand capacity in another location, Kwangyang, as the Pohang site was full (see Table 3.1).

The construction of the Kwangyang plant started in 1985, and in four phases, by 1992, POSCO had added another 11.4 million tons of annual production capacity in Kwangyang when total capacity reached 21 million tons. Since 1992, POSCO has improved plant efficiency and become the second largest steel mill in the world.[11] These subsequent expansions of POSCO's capacity can be explained by the excellent performance of existing plants supported by rapid learning and absorption of imported technology and know-how.

Performance and international competition

It is remarkable that POSCO has shown excellent financial performance ever since its operations started in 1973. However, considering the government support for POSCO during the 1973–81 period, in the form of tax incentives and construction of infrastructure such as a harbour and roads, it is difficult to argue that profits in the early period were exclusively due to POSCO's efficient management. Despite the rapidly changing business environment in Korea and the world, which has affected POSCO's performance – including increases in wages and changes in the exchange and interest rates – the growth of sales and profitability of POSCO has continued.

Table 3.2 shows the sales and profitability of POSCO during the 1989–95 period. It suggests that POSCO's sales growth was very rapid while profitability rose in the 1990s, especially as capacity expansion at the Kwangyang site was nearly completed in 1992. During the six years between 1989 and 1995, total sales grew 88.3 per cent from 4,364 billion won (US$6.5 billion) to 8,219 billion won (US$10.7 billion). The profits grew much faster by 480 per cent from 144 billion won (US$214 million) to 840 billion won (US$1,089 million) during the same period. The profit-sales ratio improved from 3.3 per cent in 1989 to 10.2 per cent in 1995, a very significant improvement in profitability. We should note that POSCO has been using an accelerated depreciation method in its accounting to increase its internal reserves, which was in turn used for financing capacity expansion during the past twenty years. If the company used a normal depreciation

Table 3.2 Financial position of POSCO, 1989–95 (billion won)

	1989	*1990*	*1991*	*1992*	*1993*	*1994*	*1995*
Sales	4,364	4,805	5,827	6,182	6,920	7,314	8,219
Profits	145	79	146	185	295	383	840

Source: POSCO, *Annual Report* (1989–95).

method, its profitability would have been higher. When we compare the financial performance of POSCO with other major international steel producers, POSCO's performance was one of the best in the world.[12]

Basically, POSCO's financial performance reflects two sets of factors: those related to factors such as yields and productivity, machinery and equipment, and organisation, which the company can control, and factors such as wages and interest rates which are given, and which the company cannot control. The remarkable performance of POSCO in the past twenty years reflects its basic competitiveness, although the business environment has been important as well. It suggests that POSCO's growing profitability in the 1990s reflects the company's long-term efforts to improve productivity and overall efficiency.

Table 3.3 shows indicators of POSCO's operational efficiency in the 1980s, compared with those of Japan, the world leader in steel making, for 1984. Nearly all the indicators show POSCO's relative inefficiency compared with those for Japan. But when we compare POSCO's data for the late 1980s with those of Japan for 1984, POSCO's performance was pretty close to Japan's. For instance, the yield ratio is almost the same; the rate of breakdown of production equipment and machinery, and the man hours to produce a ton of steel were close to the performance of Japanese steel producers. The data in Table 3.3 suggest that the growth of productivity and efficiency of POSCO has been very rapid.

Table 3.4 shows that around 1986, POSCO had already become one of the lowest cost producers among major steel producing firms in the world. The price competitiveness of POSCO seemed to be closely related to the following factors: low labour costs (US$9,277.3 per worker), high capacity utilisation (0.9927), high value-added per capita (US$68,100), and new machinery and equipment. POSCO's labour productivity was similar to that of steel producers in advanced countries, but the large differences in wage rates between Korea and these countries made POSCO's labour costs one of the lowest among the major competitors.

It is important to note that, except for four years, POSCO has continuously maintained a 100 per cent capacity utilisation ratio since its start-up in 1973.[13] The rapid and sustained growth of the Korean economy, and the efficient management and price competitiveness of POSCO made maximum utilisation of POSCO's capacity possible, which in turn reduced POSCO's costs of production further.

Table 3.5 compares costs of major products of integrated steel mills in three countries – Korea, Japan and the US – and a mini-mill (Nucor) for slab, hot-rolled

Table 3.3 POSCO's steel-making efficiency indicators

Efficiency indicators	*1979*	*1980*	*1984*	*1985*	*1986*	*1987*	*1988*	*1989*	*Japan's 1984 performance*
Yield to finished product (%)	81.8	81	88.1	88.5	89.7	91.5			92.8
Continuous casting ratio	18.2	16.7	48.8	51.3	62.4	79.4			96
Energy consumption ratio (1000 kcal/T-S)	5,785	5,515	5,385	5,356	5,205	5,205	5,228	5,194	4,696
Man hours per ton (MH/Ton)	8.15	8.12	5.51	5.34	5.22	5.04	4.72	4.49	3.04
No. of types of steel developed (cumulative)	189	205	364	399	413	426			582
Rate of breakdown of production facilities (%)	1.88	1.77	0.53	0.3	0.26	0.53			0.61

Source: POSCO, *Annual Report* (various years).

Table 3.4 Performance of major steel companies in the world, 1986

Companies	*Competition indicators*							
	Distribution ratio of wage to gross value-added	*Value-added per capita ($/man)*	*Labour cost per capita ($/man)*	*Sales per capita ($/man)*	*Shipment per capita (ton/man)*	*Sales price per ton ($/ton)*	*Equipment per capita ($/man)*	*Capacity utilisation*
POSCO	0.1362	68,100	9,277.3	174,376.0	604.8	288.33	244,146.6	0.9927
NSC	0.5887	67,000	39,440.6	201,623.9	418.6	481.69	144,116.7	0.5289
NKK	0.4418	92,800	40,999.8	217,134.5	451.8	480.61	305,092.7	0.5190
USX	1.3036	27,100	35,328.8	95,785.9	197.5	485.00	60,947.7	0.3973
Thyssen	0.7946	26,300	20,898.9	91,558.3	193.2	473.83	25,390.3	0.8450
Usiner	0.7686	36,200	27,823.0	116,861.3	227.8	513.07	53,935.6	0.5071
British	0.6423	32,000	20,554.2	93,324.2	207.1	450.60	56,023.3	0.5897
BHP	0.5698	34,800	19,828.3	66,601.6	135.3	492.21	96,260.2	0.6915
Voest	0.9876	29,100	28,737.7	65,008.8	166.3	390.94	51,000.8	0.8298
CSC	0.2102	55,300	11,622.0	109,896.5	378.5	290.35	236,974.3	0.9737

Sources: PaineWebber, *World Steel Dynamics*, 1987; RIST (1989).

Table 3.5 Comparison of production costs, 1993 ($/ton)

	USA	*Japan*	*Nucor*	*POSCO*
Slab	231	229	245	167
Hot-rolled coil	302	303	288	184
Cold-rolled coil				
raw materials	141	125	164	107
labour	156	170	42	41
other	141	186	140	114
sub-total	438	481	346	262
depreciation	26	82	39	115
financing cost	15	24	—	21
total	479	587	385	398

Source: PaineWebber, *World Steel Dynamics, 1994*.

coil and cold-rolled coil. It also reveals the cost breakdown of inputs for cold-rolled coil in the three countries around 1993. In the case of slab and hot-rolled coil, POSCO's production costs were the lowest. This is also true for cold-rolled coil, except the products of Nucor, mini-mill producer. Even in this case, the price was very close, and the difference mainly resulted from the accelerated depreciation methods adopted by POSCO. We need to add another important source of POSCO's price competitiveness, namely the low construction cost of steel plants. POSCO's construction costs were the lowest among major steel producers in the world (PaineWebber, *World Steel Dynamics, 1989*).

Table 3.6 shows selected indicators of POSCO's technological capability and competitiveness compared with those of Japan, the world leader in steel making, during the 1980s and 1990s. The ratios for continuous casting and (steel) yield, both of which have been critical for competitiveness, clearly indicate how fast POSCO was able to catch up with Japan.[14]

What are the key factors responsible for the excellent performance of POSCO? Four major factors contributed to the company's successful performance: (1) low costs of production; (2) government support; (3) the development of indigenous technological capability with heavy investments in human capital and R&D; and (4) excellent managerial capability, including operations fully utilising capacity and co-operative management–labour relations.[15] All four factors were equally important in POSCO's success. In terms of the critical factors that developing countries should consider, the development of indigenous technological capability and managerial capability should be emphasised.

First of all, low labour costs have been an oft-cited factor behind POSCO's competitiveness. The labour costs of producing a ton of cold-rolled coil at POSCO were about a quarter of Japan's or the USA's in 1993 (Table 3.5). Despite a sharp rise in wage rates in the 1987–9 period and the fast appreciation of the Korean

Table 3.6 Selected indicators of technology and competitiveness (%)

	1986 Korea	*1988 Japan*	*1989 Korea*	*1992 Japan*	*1993 Korea*	*1994 Japan*	*Korea*	*Japan*	*Korea*	*Japan*	*Korea*	*Japan*
Continuous casting ratio	75.1	92.7	88.3	93.1	94.1	93.5	96.8	95.4	97.8	95.7	97.8	—
Yield (steel)	89.7	92.6	91.7	92.6	92.6	92.8	—	—	94.6	93.1	95.2	—
Speciality steel	—	18.0	6.5	18.7	6.5	17.9	8.2	17.8	8.9	17.6	10.7	—

Sources: Japan: *Statistical Yearbook of Steel Industry*; Korea: *Cholgang tonggye yonbo* (Steel Statistical Yearbook), Seoul: Hanguk Cholgang Hyophoe.

won, POSCO's relative wage rates did not increase much compared with other major competitors internationally. This was partly because rises in wage rates were offset by labour productivity increases. The man-hours required to produce a ton of steel, for example, declined from 8.12 hours in 1980 to 4.39 hours in 1989. In the case of cold-rolled coil, the man-hours required to produce a ton of steel were reduced from 8.5 hours in 1988 to 6.1 hours in 1995. Although labour productivity was lower than Japan's (4.9 hours in 1995), the wage gap between Japan and Korea meant lower labour costs for POSCO relative to Japan.

Relatively low construction costs also helped POSCO reduce its production costs in the early period. The average construction cost per ton of crude steel at POSCO was $287 in 1973, but this increased to $637 in 1987 when POSCO completed the first phase of its Kwangyang plant. However, this was still less than half the $1,421 that it cost the US in 1985 (Barnett and Crandall, 1986). POSCO's low costs of construction can be attributed to a combination of elements: (1) efficient management of construction work by experienced Korean construction companies; (2) low labour costs; (3) low financing costs; (4) relatively short construction period (e.g. the first phase of construction of Kwangyang's 2.7 million ton capacity took only 26 months to complete, whereas Japan took 33 months to build NKK's second phase with a 3 million ton capacity); (5) relatively fast start-up

POSCO's good performance has frequently been attributed to subsidisation by the Korean government. Government subsidies helped lower production costs during the start-up period in the early 1970s. According to an estimate from the Korea Advanced Institute of Science and Technology (KAIST) in 1976, the government provided about $42 million in the early 1970s to support POSCO's start-up, mainly in the form of infrastructure building, e.g. roads and harbours, and for the provision of low-interest long-term, foreign loans to purchase equipment. Since the late 1970s, however, the subsidies have been discontinued. Since POSCO has been required by government guidelines to sell POSCO products in the domestic market below international prices, the subsidy argument cannot adequately explain POSCO's performance.

It is important to recognise excellent managerial capability as a key factor in POSCO's success. Although POSCO was set up as a state-owned enterprise, it has been able to avoid the 'usual' government intervention in its management and operations. This was possible because of the government decision to make POSCO's management independent of government intervention and red tape. POSCO's President Park was able to manage the company just like a private firm, and maximum efficiency was the guiding principle. The rapid growth of the Korean economy through industrialisation, which led to the growing demand for steel, also helped POSCO since it enabled the company to operate at nearly full capacity most of the time.

Finally, the development of indigenous technological capabilities in steel production as well as the learning and high absorption capacities of POSCO were key factors, which enabled POSCO to catch up with and threaten global leaders. From the very beginning, POSCO made plans to develop indigenous technological capabilities so that it could eventually compete with the global leaders, and massive investments in training and R&D followed. The results were rapid learning, assimilation and innovation that increased productivity and the production of new steel products. This point will be elaborated below.

The development of POSCO from its inception in 1973 until 1995 is a fascinating case, as it started from nothing (i.e. no technology, no skill, no experienced managers, no capital and no raw material) and grew to be a premier world-class steel producer. This is, indeed, the story of a 'created comparative advantage', as Amsden (1989) put it.

Development of technology and technological capability

We learnt above that the development of POSCO's technological capability brought about successful assimilation and innovation in POSCO's product and process technology, contributing importantly to POSCO's phenomenal success. We now discuss how POSCO was able to develop the technological capability to innovate firm-specific product and process technology. POSCO – which did not have any experience or indigenous technological capability in steel making – had to rely on 'learning' product and process technology from experienced steel producers and from suppliers of steel-making machinery and equipment, in which 'learning' was the first step.

Learning as the basis of innovation

POSCO engineers quickly learnt how to build and operate a modern integrated steel mill in the initial period with the help of Japanese engineers, as Korea did not have any experience in building or operating an integrated steel plant. During the first phase of construction of the steel mill, POSCO engineers worked closely with their Japanese counterparts. In addition, a large number of Korean engineers received intensive training covering all fields of iron and steel making, in overseas plant sites, mainly in Japan and Austria, before first phase plant operations began. Their efficient learning led to a smooth early start-up of the new plant. In the second year, most plants operated at rates above full nominal capacity. This was a key factor underlying how POSCO produced profit early on.

Learning by POSCO staff about plant operation and engineering as well as construction engineering accelerated as POSCO added new plants at the Pohang complex in Phase II to reach 2.6 million tons annual production capacity. During Phase II, the role of Korean engineers who had worked closely with foreign engineers became greater as their experience grew.

Extensive training of POSCO engineers and continual investment in human resources and in related facilities have been the conditions that have helped POSCO to maintain a high capacity utilisation rate and perform well from early on. By the mid-1980s, only about ten years since its start-up, POSCO's technical efficiency had reached a level close to that of Japan, the most efficient steel producer in the world, in terms of finished product yield, energy consumption and the actual rate of depreciation of production facilities. In recent years, POSCO's efficiency and productivity have improved very significantly due to intensive learning and accumulation of innovations.[16] For instance, labour productivity, measured in man-hours required to produce a ton of steel, grew over 30 per cent from 4.4 hours in 1992 to 2.9 hours in 1996. Value-added per employee grew from 106 million won to 176 million won over the same period. The capacity utilisation rate was always over 100 per cent during the 1990s, reaching 110.7 per cent in 1996. The share of high value-added steel in total production grew from 26.2 per cent in 1992 to 30.4 per cent in 1996, and the share of cold-rolled products increased from 30.4 per cent to 36.0 per cent over the same period. It is to be emphasised that innovations in POSCO have been due to learning.

The chain-linked model of innovation

We learned that a key factor in explaining the excellent performance of POSCO in the past thirty years was active innovation and product development. Although POSCO, as a latecomer in the field of integrated steel making, has not invented many new products or process in steel making, it has made many innovation efforts in steel-making technology. After all, technology and productivity growth is due to both incremental and breakthrough innovations. In recent years, many studies, especially of process technology, show that incremental innovations have been as important as breakthrough innovations (inventions) in technological and productivity growth (Utterback, 1996, chapter 6). POSCO's innovations are largely in the area of incremental innovation.

In order to systematically explain POSCO's successful innovation efforts, the chain-linked model, developed by Klein and Rosenberg (see Aoki, 1989, chapter 3), seems to be useful. Aoki has used the model to explain the innovation activities of Japanese firms in the post-war period. Although Japan has not invented many new products or processes, it has been very successful in the commercial development of many inventions through intensive innovation efforts, e.g. transistorised radio and television sets, VCRs, copiers, etc.

The Rosenberg–Aoki chain-linked model has been offered as an alternative explanation of Japanese-type innovation activities, instead of the traditional linear model. The linear model seeks to explain the innovation process for new product development from basic research through applied research, product (process) development, manufacturing and marketing, but cannot adequately explain many Japanese innovation activities. The first key element of the chain-linked model that differentiates it from the linear model is that the innovation process is not necessarily linear as many possible channels exist. Second, scientific research capability for invention is not the only source of innovation as the stock of existing knowledge is also a good source of product innovation. Third, the feedback process among the key players in an innovation process, namely the scientists, researchers, engineers, line-workers and users of products, plays a key role in the model. For instance, the improvement of a product's functions is often initiated by the market, e.g. in response to users' complaints or rival products.

Close interactions among designers who perform basic engineering and detailed engineering, manufacturing engineers, line-workers and final users of a product are important for innovation, including the development of improved products. These types of innovation process are based on continuous and interactive information exchange and co-operation among major players in the value chain. For the same reason, close interactions among specialists in related but different departments are crucial for the innovation process. Close co-operation between researchers in the research department and engineers in the production department who possess real knowledge of the manufacturing line is essential for successful innovation of this kind.

In sum, we can conclude that feedback among major players in the value chain of product development, namely researchers and designers in the research department, engineers in the production department, line-workers and final users, as well as close co-operation among them are critical elements in successful innovation, according to the chain-linked model. We will see below that POSCO's major innovations can be explained by the chain-linked model.

Evolution of POSCO's technology development system

Before we discuss the major innovation efforts of POSCO, it is necessary to study how POSCO's technology development system evolved. The evolution of POSCO's technology development system will show important elements in its successful innovation efforts.

In 1973, when the first stage of POSCO's construction was completed, the POSCO management decided to establish a research centre in order to conduct systematic research and development activities in the field of steel production. This was regarded as essential for the development of POSCO's global competitiveness. As a first step, the technology development department of the Pohang steel plant was enlarged in 1977, and in the next year, the Research Institute of Science and Technology was formally established. The refusal by major steel producers and equipment manufacturing firms to provide technology and equipment to POSCO for the construction of the Kwangyang plant provided strong motivation for POSCO to step up research and development activities of its own in order to reduce technology dependence on the leading steel producers and engineering companies. Finally, in 1987, the enlarged R&D centre was established as the Research Institute of Industrial Science and Technology (RIST).

In 1986, Pohang Institute of Science and Technology (POSTEC) was established by POSCO as a university dedicated to industrial research and graduate teaching. POSTEC is located in Pohang near POSCO and works closely with POSCO in steel related industrial R&D.

In 1994, POSCO set up in-house the POSCO Technology Centre to deal with steel-related R&D issues directly affecting POSCO's performance, as POSTEC and RIST's R&D work was oriented towards POSCO's long-term projects. In 1994 and 1995, POSCO established technology centres in Japan and Germany to monitor global steel-making technology change and to strengthen joint research and strategic alliances with leading steel R&D institutions.

Internally, POSCO established the Central Technology Office under its Chief Technology Officer (CTO) in 1996, whose main function is to co-ordinate the R&D activities of RIST, POSTEC, the in-house Technology Center and POSCO's R&D affiliates in Japan and Europe. More importantly, the office of the CTO was to play a key role in deciding on major R&D projects and the on direction of POSCO's R&D strategy. As will be shown later, close co-operation among key R&D institutions and the steel mill was crucial for POSCO's successful innovation activities.

POSCO's major innovations

Development of TMCP steel

Demand for specialty steel needed for the shipbuilding industry and the off-shore oil industry grew rapidly during the late 1980s. Some advanced steel firms in Europe and Japan developed thermo-mechanical controlled process (TMCP) steel in the early 1980s to satisfy industry demand, but POSCO had to develop its own TMCP steel-making process, since it could not get the necessary technology from abroad. Demand for TMCP steel for shipbuilding and the off-shore oil industry was due to its properties meeting industry requirements. The shipbuilding industry needed steel plate to be flexible enough to increase productivity by not requiring preheating for better welding, while the off-shore oil industry wanted high strength steel to prevent brittle fracture in deep cold water.

The characteristics of TMCP steel are as follows. First, its tensile strength is higher by 10 kg/sq mm over conventional normalised steel without the addition of any alloy element. The ductile to brittle transition temperature is decreased by 50 degrees (centigrade) due to ferrite grain refinement. Second, TMCP steel is easier to use compared with conventional steel because preheating for welding can be eliminated and high heat input welding can be done without reducing the strength of the heat-affected zone. Third, the production cost of TMCP steel is lower than for conventional steel by eliminating the need for off-line heat treatment and by reducing the alloy element.

In 1987, POSCO finally decided to develop its own TMCP steel and formed a TMCP committee consisting of research teams from four departments, namely the hot rolling and quality control departments from POSCO, the plate research department and the welding research laboratory from RIST. These teams began basic research concerning properties such as alloy design, microstructure, temperature control and plate distortions. From 1988, another team began to develop production technology for TMCP steel. By late 1988, these teams were able to get satisfactory research results, and on this basis, detailed purchase specification for an accelerated cooling facility was prepared, and the facility installed in November 1989. An eight-month trial period was necessary before full commercial production could begin in late 1990. One important element in the successful development of TMPC steel was the close co-operation among the four related departments from POSCO and RIST that formed the joint research team.

Smelting reduction method of steel making: Corex and Finex

In 1993, POSCO decided to invest in a new steel mill with an annual capacity of 600,000 tons using a new smelting reduction method called Corex, originally developed by Vost Alpine of Austria. This new technology in steel making replaced traditional blast furnace and pre-processing requirements for iron ore and coke with a smelting reduction furnace. Only South Africa had a small plant with a 300,000 ton capacity using the same technology. The mill was completed at POSCO in 1995 and by 1996, it was commercially successful with a capacity utilisation rate of 120 per cent. However, although the Corex method eliminates the coke production phase, it requires iron pellets whose production produces SO_x and NO_x, because it cannot use (pulverised) natural iron ore.

POSCO – in collaboration with RIST, POSCO's own affiliated research institute and Vost Alpine of Austria – has been developing a new smelt reduction method called Finex. Finex can use (pulverised) natural iron ore, and can eliminate the environmental problems associated with the Corex system. POSCO and RIST have nearly completed pre-commercial production tests. They have built an industrial scale plant for final testing of the Finex system. So far, the Finex system has had superior test results compared with the conventional blast furnace method in terms of preventing environmental problems, productivity, cost reduction and investment capital requirements.

The development of the Finex system was possible because POSCO and RIST have collaborated closely, and also used Vost Alpine, which has developed the Corex system, as a collaborator. First, POSCO mastered the Corex system based on the smelt reduction method by constructing a steel plant based on the Corex system before developing an advanced smelt reduction method called Finex in order to reduce risk. This is another successful case of incremental innovation efforts by a latecomer.

Strip casting process

POSCO and RIST, in co-operation with Davy Distington of the UK, have been collaborating since 1990 to develop a twin roll casting process for steel sheets. The final goal of this project is to develop a direct strip casting process for stainless and carbon steel with full commercial viability. Progress in developing the strip casting process was in two phases as two pilot plant casters were constructed. In the first phase, the first caster – capable of producing steel sheets of a width of 350 mm with a thickness between 2 and 6 mm – was installed in POSCO's Pohang plant in 1991, with the first full cast of one ton made in December 1991. After many trials with the first caster, POSCO designed and built the second caster in 1995 for a width of 1,300 mm at a thickness of 2 to 6 mm with a ten ton capacity.

In addition, the thin slab casting method was developed in 1996 in POSCO Kwangyang's mini-mill plant, raising productivity in steel making by over 20 per cent by eliminating the crude rolling process. POSCO has been improving the thin slab casting method for installation in its second mini-mill in Kwangyang to be completed in 1999.

Utilisation of information technology at the Kwangyang plant

Since the Kwangyang plant (KP) was the later addition among the POSCO plants, the information technology (IT) used for the KP involved the most advanced steel manufacturing system in Korea. This section will explain how the KP system was planned, developed and implemented.

The Kwangyang plant was designed to maximise operating efficiency with several features to achieve this goal including: (1) a compact layout of facilities to raise the efficiency of material flow and to minimise handling costs; (2) 100 per cent continuous casting and a direct linkage between the steel making and hot-strip rolling processes with the hot charge and rolling (HCR) system for cost reduction and energy conservation; (3) an integrated IT system for all production processes from the time a customer order is received to the time of shipment. Advanced iron- and steel-making technologies available commercially were adopted to improve productivity and quality.

The 'real-time production control system' adopted at the KP included the following features: (1) a total system for integrating control functions from upstream through rolling to shipment; (2) a high speed network to link the business computer with all the process computers and terminals; (3) organisation of the database to make it accessible at any point for any application; (4) an efficient system that enables non-stop operations for 24 hours a day, with the ability for fast back-up and recovery.

The application software configuration handles quality control, shipping, operation, energy, maintenance, accounting, revenue and virtually all aspects of production and business at the KP. Some key systems for production operations include the following: (1) a system for the scheduling of material flows, named HIPASS; (2) an on-line quality evaluation system; (3) a slab yard control system; (4) a system for shipping; (5) a process control computer system. A brief description of important features of each system follows.

1 Operation of the 'hot charge integrated process adjusting and scheduling system' (HIPASS) begins with plans received from the production planning system. The key functions of HIPASS are time scheduling, standardisation and condition control, work instruction, real-time tracking and monitoring, emergency management and reporting of the work results.

   HIPASS can handle production scheduling for up to five days. The original manufacturing plans are sometimes rescheduled when modifications are needed. Rescheduling takes place several times a day on average. Time scheduling is done either by the Blast Furnace-BOF Scheduler or the BOF-Hot

Strip Mill Scheduler. Both schedulers have eight scheduling packages and five rescheduling packages. The packages run either independently or together, depending on how scheduling and rescheduling are performed. The forward and backward simulation of the actual manufacturing process can be done to help scheduling. The scheduling simulation is done in detailed basic units instead of large production lots, so that minute-by-minute loading schedules for all facilities can be shown. The typical scheduling time required for 100 heats of BOF operations is one to three minutes for both schedulers on the current host computer. Information on the materials flows is monitored through the audio and video terminals installed in the production control room.

2 The Quality Evaluation System (QES) is used to analyse data of the 156 variables which affect slab quality collected from process computers and sensors during steel making, refining and casting. A model is used to evaluate the data and classify them to form four quality characteristics for slab surface quality and six for slab internal quality. These characteristics are compared with the design criteria and the quality of the product is judged. The quality evaluation system is used to improve quality and to reduce defects by continuously reanalysing data for the 156 variables in the data bank to predict the occurrence of defects.

3 The main functions of the slab yard control system (SYCS) consist of the following processes: slab destination control, automated location control and charge sequence control of the re-heating furnace. The real-time yard status map is maintained by slab tracking. Through the SYCS, the Kwangyang plant saves time and thus speeds up the preparatory process for rolling production.

4 Warehouse control and pallet transportation control are conducted through the shipping system (SS). The SS warehouse control function locates and selects coils to be shipped according to the vessel storage plan while the pallet transportation control is automatic and tracks pallet transportation. The new SS enhanced efficiency by saving time from warehouse to vessel, and reducing the required manpower.

5 The process computer system, together with lower level PLCs and instrumentation equipment, are the main equipment items for plant automation. The main functions of the process computer system are process control, data gathering and operation guidance linking with the business computer.

Project management

Setting up the real-time production control system at the Kwangyang plant was difficult, due to the complexity of interfacing a number of systems vertically and horizontally. The management was constrained by the time limit and the lack of experienced personnel. The project task was divided into four groups: application development group, progress control office, test administration team and development support group.

The application development group is divided into nine teams according to function. Their respective functions included: order entry and shipping planning, production planning, progress control, HIPASS, operation control, data bank, energy management, facilities maintenance and general administration. Each team consists of two groups, end-users who use developed systems and system designers. In the basic design stage, end-users took the initiative and led the team. In the detailed design and programming stage, system designers played the key role. At the testing and operation stage, the end-users again took the lead to determine whether the developed system worked as designed.

The main function of the progress control office (PCO) is to co-ordinate each task and the work progress of each team, to optimise utilisation of the overall system developed under time constraints and conditions of insufficient supply of experienced personnel.

The test administration team was set up to perform tests at each stage of development and to review their results. The test was done at the detailed design stage as well as at the unit test stage. The software developed was tested programme by programme to achieve an integrated overall system with test scenarios and simulation data.

System development history

The total time required to develop the production control system at the Kwangyang plant was 7,100 man-months. For phase I of the KP, it took 2.5 years to develop the system, but for phase II, it only took a year. In phase I, 8,556 programmes were developed, including 3,480 programmes developed by vendors of equipment sold to POSCO. In phase II, 8,701 programmes were developed, including 1,658 programmes by vendors.

The KP production control system incorporates all necessary control functions in a single hierarchical system, from intelligent sensors and instruments installed with machines through a PC and process computer system to the integrated management and planning system.

The main features of the Kwangyang system are real time process control, real time data base management for material flow control, a common data base for the whole Kwangyang plant, including management and planning applications, a single control centre for the whole plant, and automation of control process.

The main factors which helped POSCO to install an efficient production control system at Kwangyang plant can be summarised as follows: (1) a well-defined development plan at the beginning; (2) flexibility in adopting new methods and technologies; (3) active end-user participation at every stage in the development of the system; (4) much attention from the top management.

All these innovations are not the breakthrough type (i.e. involving original invention), but rather, are incremental innovations built on innovations by other firms. POSCO was successful in the commercial application of these foreign technologies through intensive learning and innovation efforts. These incremental innovations have been the cornerstone of POSCO's ascendance to the top among

Table 3.7 Comparison of steel product quality and process technology

	Korea	*Japan*	*Others**
Product quality			
Steel plate for auto (t.s, kg/sq mm)	80	120	50
Steel plate for shipbuilding (t.s, kg/sq mm)	50	50	40
Process technology			
Labour productivity (t/man)	866	1,102	603
Energy efficiency (1,000kcal/t)	529	589	680

Source: RIST (1989).
Note: * Others – estimates of steel plants in developing countries.

global steel producers. Table 3.7 compares steel product quality and efficiency in terms of labour productivity and energy efficiency between Korea and Japan. It shows how fast POSCO could catch up with Japan through massive learning and innovation efforts.

Factors important for technological capability building

We can now summarise the key factors behind technological capability development at POSCO, which enabled the company to learn fast and accumulate innovations.

Active participation with foreign counterparts

Since Koreans did not have experience in building and operating an integrated steel mill, initially, the whole task of preparing for the overall engineering master plan and the construction of individual plants, including basic design and start-up, was given to foreign companies, mostly Japanese engineering firms. However, Korean engineers from POSCO and other participating firms actively worked with their foreign counterparts in all engineering tasks in order to learn quickly. In addition to the instruction given to POSCO engineers concerning the basic operations of the integrated steel mill, Korean engineers accumulated expertise in areas such as production scheduling, maintenance, inventory control and benchmarking the most efficient foreign steel producers in advanced countries. In subsequent phases of expansion, the role of foreign engineers was progressively reduced, and by Phase IV, the Korean engineers were able to prepare the engineering master plan.

Overseas training and intensive in-house training

Since there were no integrated steel mills in the country, a large number of key staff, engineers and supervisors were sent abroad for intensive training. Many of them were sent even before POSCO's operation commenced. Upon returning, those who were trained abroad passed on their knowledge to their colleagues through in-house training programmes, which continued until recently. Engineers and operators were subjected to intensive in-house training to keep them abreast of the latest technological developments. Such intensive in-house training led to numerous improvements in operations and quality control. The accumulation of small changes significantly improved POSCO's competitiveness.

Research and development

Since the tasks of producing different varieties of high-quality steel as well as reducing production costs through quality improvements are important, R&D was regarded as a vital priority for the company. POSCO established a R&D centre in the 1970s, and spent more than 0.5 per cent of total sales on R&D. In the 1990s, R&D expenditure increased to more than 1 per cent of sales with the R&D centre expanded even further. To achieve long-run competitiveness through R&D and training, POSCO established an engineering college and graduate school, one of the best in Korea. As shown in Table 3.8, the R&D investments of POSCO grew very rapidly. In 1995 and 1996, the ratio of R&D investment to sales reached nearly 2 per cent, implying very rapid growth of R&D investment since the increase in sales during the 1990s. The number of scientists and engineers working as R&D staff has also grown rapidly with the total number of R&D staff reaching 1,106, senior research staff numbering 552 and 206 holding PhDs in 1996. The research department opened two overseas offices, one in Japan and another in Europe, to facilitate monitoring foreign R&D trends and to promote joint research.

Diversified sources of foreign technology

Although the main source of technology and know-how was Japanese engineering firms, POSCO actively sought other sources as well, including Austrian and German steel engineering firms. When Japan hesitated to provide advanced technology to Korea around the time the Kwangyang plant was being built, POSCO was able to move ahead without the active participation of Japanese, or even non-Japanese firms, by relying on its own technological capability.

Table 3.8 Research and development (R&D) investment

	1989	*1990*	*1991*	*1992*	*1993*	*1994*	*1995*	*1996*
Total sales (billion won)	4,364.3	4,805.0	5,827.4	6,182.1	6,920.9	7,314.0	8,218.7	8,445.5
R&D ratio (%)	0.94	0.83	0.96	1.52	1.42	1.18	1.96	1.95

Source: POSCO, *Annual Report* (1989–96).

Foreign competition

Despite government protection in the early period with a 25 per cent import duty, the company always had to face foreign competition in both international and domestic markets. Since POSCO products were intermediate goods for steel utilising industries such as the shipbuilding, automobile and machinery industries that had to compete in international markets, the Korean government and POSCO knew how important it was for POSCO to supply its products at internationally competitive prices in the domestic market. The Korean duty drawback system ensured that all exporters using imported steel could get their duty back when the finished products were exported. In addition, POSCO exported roughly a third of its products on average through the years to maintain full capacity utilisation. The competition which POSCO had to face from foreign producers in Japan, Taiwan and Brazil forced the company to learn fast and to constantly innovate to enhance competitiveness.

POSCO understands very well that international competitiveness depends on the vintage of the machinery and equipment which embody the new steel-making technology. This is why POSCO has been diligently investing in new technologies including the new smelting reduction method (Corex), strip casting and the development of new high value-added steel products. POSCO's experience in the development of technology and technological capability has clearly shown that learning is the basis of innovation, and the accumulation of small innovations based on learning is the key factor in rapidly catching up with world leaders to make POSCO one of the most efficient producers of steel in the world. From this point of view, innovation can be understood as 'a product or process new to the firm, rather than to the world' (Hobday, 2000).

Role of government

Government intervention in the market may be justified by the existence of market failure if the intervention can correct the market failure. Monopoly, externalities, public goods and asymmetries of information involve market failures. In developing countries, other reasons may justify government intervention in the form of a state-owned enterprise (SOE). First of all, markets, especially financial markets, are often weak and poorly developed. Moreover, the private sector in developing countries is generally weak, and entrepreneurship not well developed. Entrepreneurship and the private sector in developing countries are hence generally too weak to start a risky large project such as a modern integrated steel mill. Reliance on foreign sources may be considered necessary for the supply of capital and technology to start such a project. In such circumstances, the SOE may be preferred to the private sector, especially since the government has to give monopoly status to the project to achieve scale economies. The need to protect a project initially for infant industry considerations may also favour the SOE vehicle. As in the case of POSCO, the SOE form does not necessarily require government management of the firm. An SOE can be managed as a private sector firm, although owned by the state.

Reasons for establishing POSCO as a state-owned enterprise

The Korean government generally relied on the private sector to start most new import substitution projects – in shipbuilding, machinery, automobiles and so on – except for POSCO. Such projects were generally initiated by the private sector with government support, especially through fiscal and financial incentives. Why was the case of POSCO different? Why was it considered necessary to start and then maintain POSCO as a state-owned company?

The Korean economic environment when POSCO was established in 1968 was considered not suitable for private sector establishment of a large integrated steel mill due to the lack of adequate private Korean capital and technology. Even the size of domestic demand was not sufficient. POSCO was established as part of a strategy to promote heavy industrialisation by encouraging important steel-using industries such as shipbuilding, automobile and machinery. The large size of the investment as well as the high risk and uncertainty would preclude a private sector firm from investing in the project. For the government, public ownership was the only choice considering the project's various circumstances. Also, other developing countries, such as Brazil and Taiwan, had also started their integrated steel mill projects through SOEs.

Important differences between POSCO and Korea's other large import-substitution industries therefore seem related to the following factors. First, the POSCO project was the single largest investment project Korea had ever had. For instance, total investment in the Pohang integrated steel mill was 5,554 billion won, while the Kwangyang mill cost 8,459 billion won. Hence, the size of the investments were too large and risky for private firms to invest in. In fact, no one wanted to invest.[17] Second, steel production supplies raw materials and intermediate goods for many key industries in Korea, including shipbuilding, automobile, machinery, electronics and construction. Private ownership of a large integrated steel mill with monopoly status could cause serious problems, especially when the start-up costs of the new mill were largely borne by the government and government-controlled financial institutions. Finally, the government thought it important for POSCO to supply excellent steel products to domestic steel users at competitive prices in order to help Korean steel-using firms become internationally competitive. It appears that the government wanted POSCO as a state-owned company, since a profit-maximising private company would not have been able to fulfil these expectations. These three factors seem to have been important additional reasons beyond reasons related to the size of the project, the risks involved, the lack of well-developed markets, private sector weaknesses as well as managerial and entrepreneurial capabilities in Korea in the 1960s.

Government intervention in POSCO operations has raised a few issues. First, until recently, the government has helped POSCO by discouraging the establishment of other integrated steel mills. This restriction, however, was not important since private firms were not interested then. Second, steel product prices in Korea were regulated by the government, which wanted to supply steel products at internationally competitive prices. Most of the time, the Korean prices of steel products were lower than international prices, thus helping the emergence and development of steel-utilising firms and industries. However, the profitability of the company was not adversely affected by these price restrictions due to its excellent performance and efficiency.

Government support

Government support for POSCO was especially concentrated during the start-up phase (1970–3). The most important support for POSCO in this period was the mobilisation of investment funds. Without government involvement, the necessary capital to invest in the POSCO project would not have been mobilised. Subscription by the government and government-controlled banks constituted 33.2 per cent of the total investment. The government also invested US$77.2 million in the POSCO project, given to Korea by Japan as a grant to compensate for its occupation of Korea for thirty-five years from 1910. Other funds were mobilised in the form of commercial loans from Japan (37.4 per cent), Austria (13.7 per cent) and Korean banks (7.7 per cent). In addition, tax reductions and government construction of basic infrastructure, such as harbour facilities, water and roads, were important. Such support was discontinued in the early 1980s, except for some infrastructure support, which continued in the 1980s as new plants in the Kwangyang steelworks started in 1986.

During the second period (1974–81), government support was given in the form of low-interest loans to reduce POSCO's financing costs and by forgoing dividend income from POSCO to help the company increase its internal reserves through a 'no dividends' policy until 1982. This support was important in making the company profitable from early on, and to enable it to finance its own expansion projects from its reserves and profits. As a result, POSCO's reliance on outside financing for its (continual) expansion until 1991 was limited to 39 per cent of the total required funds. Of the total investment, 61 per cent, namely 1.1 trillion won out of 1.8 trillion won, was from POSCO's own reserves. The level of government support in the third period since 1982 has been appreciably lower, and basically limited to infrastructural support for new plant facilities in the Kwangyang steelworks by preparing the harbour, water supply, roads and so on.

It is not easy to evaluate the impact of government support on the performance of POSCO since it is difficult to separate the effects of government support from the effects of efficient firm management, although there is no question that government support was essential during the start-up phase of the firm. The government policy to regulate the domestic prices of POSCO products – so that the prices of POSCO products to domestic steel users would be lower than the price of imported steel products – makes it difficult to evaluate the net effect of government support for POSCO. Government regulation of the prices was justified by POSCO's status as a state-owned enterprise which had received government support (especially during the firm's start-up phase) and its monopoly status for many POSCO products in the domestic market. Although government support has been discontinued, government price control continues to the present. For example, in 1994, POSCO's loss of revenue due to price controls by the government amounted to 540 billion won (approximately US$670 million) – based on the differences between POSCO prices and the best prices for equivalent imported steel products, and the domestic sales volume of the POSCO products. In sum, except for the initial period (1973–81), POSCO's excellent performance has mainly been due to the efficient management and international competitiveness of the firm, rather than government support since most of the support had been discontinued by the early 1980s.

Government control and POSCO's evolution

The nature of government control of POSCO and the evolution of POSCO's status as a state-owned monopoly shaped the relationship between POSCO and the government. The government controlled POSCO by means of its power as the major stockholder to appoint its chief executive officer. The government also provided general guidelines for POSCO operations to ensure that the firm served the needs of national economic development and Korean steel buyers. POSCO was expected to serve local customers by providing quality products at internationally competitive prices. Except for these general guidelines, POSCO operated like a typical private sector firm, rather than an SOE. For instance, POSCO had complete managerial autonomy in terms of setting management objectives, managerial decision making and appointment of officers. POSCO did not enjoy any special privileges as an SOE in terms of getting government support besides what has been described earlier. Legally, POSCO is a profit maximising company under the Commercial Law of Korea and pays taxes at the normal rate of 34 per cent of pre-tax income just like any other private sector company, whereas a typical SOE pays tax at a reduced rate of 25 per cent of pre-tax income. Since 1988, when the government decided to sell POSCO stock to the general public and to employees of the company, the share held by the government and government-controlled Korea Development Bank has been reduced from 69.1 per cent to 33.7 per cent. Since 1992, foreigners have also been allowed to buy POSCO stock. The government still holds a controlling share in POSCO, but the company enjoys complete managerial autonomy under the broad general guidelines of the government in order to serve the national economy and Korean firms, as in the past.

The government's recognition of POSCO's monopoly status is significant. In recent years, POSCO has supplied about 50 per cent of total domestic steel demand on average. POSCO was designated by the government as a market controlling producer of five products, including hot- and cold-rolled coil, under the Korean Fair Trade Law. However, POSCO has been competing with foreign and domestic suppliers/producers in domestic and foreign markets. Therefore, if the company's competitiveness declines, POSCO's market share will certainly be reduced. Although the government holds a controlling share in POSCO, POSCO's excellent performance comes from its efficiency and international competitiveness rather than its relationship with the government. Hence, we can conclude that the Korean government – to the benefit of the national economy – has enjoyed high rates of return on its investments in POSCO.

Concluding remarks

The case of POSCO is a fascinating example of how a developing country that initially did not have any significant factor inputs, such as capital, experienced and skilled labour, or technology, built a modern integrated steel mill and made it a world-class steel mill in terms of size, quality and performance within two decades of operations. Another interesting fact is that POSCO was started as and stayed a state-owned firm.

It is clear that the Korean government, which conceived the idea of an integrated steel mill, played a critical role in making POSCO a success story, initially through subsidies and other support. However, it is important to note that government intervention was minimal, and the company was managed just like a privately-owned firm. The government required POSCO to be efficient in the face of international competition, initially by supplying POSCO products in the domestic market at prices lower than the international prices. In that sense, SOE status and early support for the firm by the government was not free. Considering the underdevelopment of the financial market and the private sector in the 1960s, government intervention appears to have been critical.

Our evidence has shown that learning and innovation have been the basis for POSCO's excellent performance, and rapid learning was the basis for its innovation. POSCO's success was thus due to technological capability building. Although growing investments in physical and human capital were important, without learning and innovation the development of POSCO into a world-class producer of steel would never have happened. In that sense, the argument that the rapid and sustained growth of NIEs has been primarily due to the rapid growth of factor inputs, rather than innovation, cannot be supported.

An interesting question is whether other developing countries should copy the Korean approach to create new industries and to sustain long-term growth through industrialisation. The answer to this question is not simple. What has happened in the case of the former Soviet Union and other countries where factor inputs were available, but could not sustain growth, suggests that factor availability was not enough to ensure sustained long-term growth. What, then, should a developing country have in addition to the critical investment inputs in order to sustain long-term growth through industrialisation?

Differences in growth performance cannot be simply explained by differences in factor supplies. How such factors are managed and utilised may be as important as or even more significant than differences in factor inputs. The concept of social capability comes from the notion that there are certain social elements which may make certain factors more productive than would otherwise be the case. The observation that certain elements in a society may have a significant effect on the productivity of particular factors is certainly not new. John Stuart Mill, Gunnar Myrdal, Simon Kuznets, and, more recently, the new institutional economics have all recognised this for a long time.

While the availability of physical and human capital are undoubtedly key to explaining the varied performances of developing economies in the past few decades, a country's capability in managing and utilising physical and human capital to make them more productive is often critical. To put it differently, why can some economies build and maintain a system which attracts physical and human capital and utilise them efficiently while others experience capital flight and brain drain? Also, why do some economic systems create positive externalities, for instance, through appropriate specialisation, communication and interaction among specialists, all of whom are important for productivity growth, while other systems fail to do so. The former USSR or India must have had much more well-trained and skilled human resources than Singapore or Hong Kong, but the management and utilisation of the human resources in these two regions was rather different. In conducive environments such as the Silicon Valley and some well-run industrial and science parks, engineers work, interact and communicate freely, learning and innovating all the time, while they often fail to do so in less encouraging contexts.

Many conventional economists neglect analysis of the environment in which production and exchange takes place by assuming that agents, individuals, firms and the government have complete and perfect information concerning products and markets over which there is no uncertainty. Hence, in such an environment, buyers and sellers will produce and exchange smoothly with no transaction costs involved. Hence, no attention is given to the economic and social milieu in which production and exchange take place.

We suggest that a variable, which we call social capability (Koo and Perkins, 1995), is a critical factor for explaining long-term performance differences in economic growth besides factor inputs such as capital and labour. As already noted, social capability is understood here as the capability of a society to manage scarce resources, but what do we mean by this? While the availability of factor inputs, such as labour and capital, are important, how effectively and efficiently these factor inputs are utilised, organised and managed for economic growth is crucial. Certain social systems and other critical features of a society render certain factor inputs more or less productive. We suggest that such managerial capability in an economy is a key explanatory variable accounting for significant differences in long-term economic growth.

Social capability is made up of several factors. First, it is composed of certain elements such as social norms, values and networks which may 'improve [the] efficiency of a society by facilitating co-ordinated action' (Putnam, 1993). Second, certain systems are crucial for maintaining a viable society, such as education, health, financial, legal and innovation systems. Some characteristics of these ingredients of social capability seem to be common: (1) they are relatively indigenous to society, and not easily brought in from outside; (2) they influence the behaviour of all economic agents (people), organisations and the governments; (3) they can be nurtured, developed and changed, but it takes time to learn and to change. Social capability is also related to culture and history, rendering 'path dependence' important. If neglected, it can be weakened and even destroyed. Social capability may help accentuate a virtuous circle if nurtured, and exacerbate a vicious circle if neglected; (4) it has the characteristics of public good. Hence, the successes of Korea and other NIEs may only be emulated by other developing countries if they develop comparable social capabilities.

The main findings of this chapter can be summarised as follows: first, the role of government was critical for both the setting up and then the successful operations of POSCO. Without government intervention in the initial period, POSCO would not have been successful. POSCO's success and other similar cases suggests that the key analytical question to ask is not whether government intervention works, but under what circumstances it works, or does not work. The POSCO management's autonomy from government control and the competition pressures that the company has faced in both domestic and foreign markets have also been important disciplining factors. Second, rapid learning has been another critical factor explaining POSCO's success as learning has been the basis for successful innovation. The accumulation of small innovations has been important in POSCO's efforts to catch up with Japan, the world leader in steel making, in both product and process technologies. The acquisition, adaptation and improvement of imported technology and related innovations have been crucial for POSCO's technology development.

Acknowledgements

For this study, I am grateful to many people who helped me with all the necessary information concerning the development of POSCO. In particular, Mr Yong-Sun Chough, former vice president of POSCO, was most helpful. In addition, I should mention that an earlier study of POSCO by Alice Amsden (1989) was very useful for this study.

Notes

1 The World Bank (1993) study, *The East Asian Miracle*, was an important contribution, although written from a neo-classical point of view.
2 For the market friendly approach, see World Bank (1991), *World Development Report*.
3 Empirical studies by Kim (1977) about the development of Korea's industrial technology have shown a close relationship between innovation and learning.
4 For the importance of organisational learning, see Kim (1977: 96–7).
5 The main reason why the export-oriented industrialisation strategy has been more efficiency-inducing than the import-substitution industrialisation strategy is that the former forces firms to face international competition, whereas the latter protects firms from such competition.
6 How governments facilitated technological development in Asian NIEs can be seen in the following: for Korea, see Kim (1977); for Taiwan, see Dahlman and Sananikone (1990), for Singapore, see Wong Poh-Kam (2001a, b).
7 For an excellent account of the development of POSCO, see POSCO (1993).
8 The opposition by the World Bank to the government plan to build an integrated steel mill was critical. See POSCO (1993).
9 POSCO's investments in the Pohang site was approximately US$9.1 billion, and in Kwangyang around US$10.8 billion.
10 POSCO, as a company under the Commercial Law of Korea, has been paying taxes at the rate of 34 per cent instead of the usual 25 per cent rate applied to state-owned public companies.

11 POSCO was expanding its annual capacity by 3 million tons at the Kwangyang site. By the time it was completed in September 1998, POSCO became the world's largest steel mill.
12 For comparative data, see various issues of *World Steel Dynamics* published by PaineWebber.
13 In fact, in most years, the real capacity utilisation rate of POSCO was over 100 per cent.
14 The continuous casting (CC) method was an important innovation for improving yield ratio (through the significant reduction of scrap) and energy efficiency, and product quality. It was only widely adopted in the 1980s. POSCO first fully adopted continuous casting throughout its manufacturing facilities in Kwangyang, and then replaced Pohang's casting facilities with the CC method.
15 In addition to the four factors mentioned, the excellent management leadership of T.J. Park – the first President of POSCO, who made POSCO an efficient learning organisation – was an important contributing factor.
16 All the data on POSCO's performance used in this chapter are from *POSCO News*, 22 May 1997.
17 The lack of interest on the part of the Korean private sector to invest in an integrated steel mill continued until the mid-1990s, when Hyundai, one of the *chaebols*, showed interest in building one.

References

Amsden, A. (1989) *Asia's Next Giant: South Korea and Late Industrialization*, New York: Oxford University Press.

Aoki, M. (1989) *The Organization and Information of Japanese Firms*, Tokyo: Toyokeizai.

Aoki, M., K. Murdock and M. Okuno-Fujiwara (1995) 'Beyond the East Asian miracle: introducing the market enhancing view,' EDI Working Paper, World Bank.

Barnett, D.F. and R.W. Crandall (1986) *Up from the Ashes: The Growth of the US Minimill Steel Sector*, Washington, DC: Brookings Institute.

Dahlman, C.J., B. Ross-Larson and L. Westphal (1987) 'Managing technological development: lessons from newly industrializing countries', *World Development* 15(6).

Dahlman, C.J. and O. Sananikone (1990) 'Technology strategy in the economy of Taiwan: exploiting foreign linkages and investing in local capability', mimeo, World Bank, Washington, DC.

Hobday, Mike (2000) 'East vs. South East Asian innovation systems: comparing OEM with TNC-led growth in electronics', in Linsu Kim and Richard Nelson (eds) *Technology, Learning and Innovation: Experiences of Newly Industrialising Economies*, Cambridge: Cambridge University Press.

Kim, Linsu (1977) *Imitation to Innovation: The Dynamics of Korea's Technological Learning*, Boston, MA: Harvard Business School Press.

Koo, B.H. and D. Perkins (eds) (1995) *Social Capability and Long-Term Economic Growth*, London: Macmillan.

Krugman, P. (1994) 'The myth of Asia's miracle', *Foreign Affairs*, 73(6): 62–78.

Lall, S. (2000) 'Technological change and industrialization in the Asian NIEs: achievements and challenges', in Linsu Kim and Richard Nelson (eds) *Technology, Learning and Innovation: Experiences of Newly Industrialising Economies*, Cambridge: Cambridge University Press.

Lynn, L.H. (1982) *How Japan Innovates: A Comparison with the US in the Case of Oxygen Steel Making*, Boulder, CO: Westview Press.

Nelson, R. (1996) *The Sources of Economic Growth*, Cambridge, MA: Harvard University Press.
Nelson, R. and H. Pack (1995) 'The Asian growth miracle and modern growth theory', mimeo, September School of Public and International Affairs, Columbia University.
Nelson, R. and S.G. Winter (1982) *An Evolutionary Theory of Economic Change*, Cambridge, MA: Harvard University Press.
PaineWebber (1985) *POSCO: Korea's Emerging Steel Giant*, World Steel Dynamics Core Report, V, New York: PaineWebber.
—— (various years) *World Steel Dynamics*, New York: PaineWebber.
POSCO (1993) *The 25 Year History of POSCO*, vol. 2: *History of Technological Development of POSCO*, POSCO.
POSCO Research Institute (1995/1996) *World Steel Handbook*, Seoul, Korea: POSRI.
Putnam, Robert (1993) *Making Democracy Work: Civic Traditions in Modern Italy*, Princeton, NJ: Princeton University Press.
RIST (1989) *Competitiveness of Steel Technology*, Seoul: Research Institute of Science and Technology.
Utterback, J.M. (1996) *Mastering the Dynamics of Innovation*, Boston, MA: Harvard Business School Press.
Wade, R. (1990) *Governing the Market: Economic Theory and the Role of Government in East Asian Industrialization*, Princeton, NJ: Princeton University Press.
Westphal, L.E., L. Kim and C.J. Dahlman (1985) 'Reflections on the Republic of Korea's acquisition of technological capability', in N. Rosenberg and C. Frischtak (eds) *International Transfer of Technology: Concepts, Measures and Comparisons*, New York: Praeger.
Wong Poh Kam (2001a) 'The role of the state in Singapore's economic development', in P.K. Wong and C.Y. Ng (eds) *Industrial Policy, Innovation and Economic Growth: The Experience of Japan and the Asian NIEs*, Singapore: Singapore University Press.
—— (2001b) 'Singapore: electronics and shipbuilding/repair industry case studies', in P.K. Wong and C.Y. Ng (eds) *Industrial Policy, Innovation and Economic Growth: The Experience of Japan and the Asian NIEs*, Singapore: Singapore University Press.
World Bank (1991) *World Development Report*, Washington, DC: World Bank.
—— (1993) *The East Asian Miracle: Economic Growth and Public Policy*, New York: Oxford University Press.
Young, A. (1995) 'The tyranny of numbers: confronting the statistical realities of the East Asian growth experience', *Quarterly Journal of Economics*, 110.

4 Taiwan's information technology industry

Wu Rong-I and Tseng Ming-Shen

The outstanding performance of the Taiwan economy during the past four decades is well documented (Galenson, 1979, 1985; Tsiang and Wu, 1985; Lau, 1990; Schive, 1990; Tsiang, 1984). The basis of Taiwan's economic development and the elements that contributed to it have been the focus of attention. Numerous theories and institutional factors have been advanced to explain the achievements of Taiwan's 'economic miracle'. With a few exceptions, almost all these efforts concentrate on macro-analysis, in which trade-related policy has been most prominent. This chapter re-examines the development of Taiwan's economy and tries to provide some new insights into Taiwan's success story by using an institutional approach to see if the government has been instrumental in enhancing private sector competitiveness by looking at the rapid development of Taiwan's information technology industry in the past ten to fifteen years.

Among the factors commonly invoked to explain Taiwan's economic development, the role of government has been the most controversial. There are two polar views in the literature: the market-friendly view and the developmental state view. The market-friendly view asserts that the economy has been very successful only because the government's role has been minimal and limited to market-oriented policies (Ranis, 1979; Tsiang, 1984). The developmental state view emphasises the leading role of the government, suggesting that it has played a central role in developing the economy by implementing various industry-promoting measures and policies. Thus, a government may substitute market forces and become a 'visible hand' helping certain industries believed to be strategically important for economic development to grow (Wade, 1990).

There has long been interest in identifying some common factors considered to have been conducive to rapid growth in the East Asian economies (World Bank, 1993; Aoki, *et al.* 1996). Instead of viewing the government and the market as mutually exclusive of each other, a third view of the role of government suggests that some policies can be adopted to make market forces work more effectively so that an economy can be developed more quickly. The idea is that the government can be market-enhancing because it can help the private sector solve co-ordination problems and market imperfections including failures. This market-enhancing view of government compensates for the fact of market failure in a developing economy, as does the developmental state view, but provides solutions which are more market-friendly.

According to this market-enhancing view, some important institutional factors conducive to rapid industrialisation in the East Asian economies have been largely ignored or misinterpreted (Deyo, 1987; Kim and Lau, 1994; Aoki, 1996). Aoki and colleagues (1996) observe that the East Asian governments have implemented their economic development policies more effectively than other developing countries because some indigenous institutions, both formal and informal, have been established to facilitate co-ordination of collective initiatives from the private sector.

This chapter uses Taiwan's information technology industry to re-examine the role of the government and to single out some important institutional elements that have contributed to the outstanding performance of the integrated circuits (ICs) and personal computer (PC) industries. The chapter is organised as follows. The next two sections briefly review Taiwan's economic development over the past forty years and the fast growth of Taiwan's information technology industry in the past fifteen years. It then introduces two important institutions critical for the development of Taiwan's information technology industry. Other important factors are discussed in the following section before the conclusion.

Taiwan's economic development policies

Over the past forty years, Taiwan has successfully and rapidly transformed its previously agriculture-dominated economy into a sophisticated global manufacturing powerhouse. By 1996, its gross national product (GNP) exceeded US$275 billion, with a per capita income over US$13,000. The average annual growth rate from 1962 to 1996 was 8.7 per cent (see Table 4.1). The sustained and rapid economic growth has made it possible for the people to increase their average income from less than US$200 in 1951 to over US$13,000 in 1996. More importantly, the rapid economic development was accomplished with relatively stable prices and very low unemployment. Since the economy took off in the 1960s, the inflation rate has been kept below 5 per cent and the unemployment rate under 3 per cent for most of the time.[1]

Another unique feature of Taiwan's economic development was that the rapidly accumulated wealth was quite evenly distributed among the people. In 1951, the income of the richest quintile was about 15 times that of the poorest quintile. This ratio dropped dramatically to approximately 5 in the 1960s and has been kept in the 4 to 5 range thereafter.

The industrial structure of Taiwan's economy has also undergone tremendous change during the past forty years (see Table 4.2). In 1961, the service and agricultural sectors accounted for 51 per cent and 25 per cent of total GDP respectively, while the industrial sector contributed 24 per cent (16 per cent from manufacturing). Over the years, the importance of industrial production has increased sharply. The contribution of industry reached the highest level in 1986, when its share of the gross domestic product (GDP) reached 45.9 per cent, and the ratio of manufacturing production to the GDP was 38.3 per cent. The manufacturing sector was clearly the locomotive of rapid economic growth in Taiwan at that

Table 4.1 Taiwan's economic growth rate (%)

Year	*Real GDP*	*Agricultural production*	*Industrial production*	*Manufacturing production*	*Service production*	*Exports*	*Imports*
1962	7.90	2.95	10.15	9.47	9.27	11.79	–5.59
1963	9.35	2.40	12.46	13.65	11.09	52.29	19.08
1964	12.20	13.15	17.27	20.69	9.41	30.42	18.23
1965	11.13	7.83	13.23	14.18	11.52	3.93	29.91
1966	8.91	2.66	14.13	17.58	8.88	19.11	11.87
1967	10.71	5.18	15.34	16.10	10.43	19.59	29.58
1968	9.17	4.93	15.36	17.15	7.30	23.09	12.03
1969	8.95	–3.97	17.65	20.38	8.49	32.95	34.33
1970	11.37	4.27	17.12	20.50	9.94	41.18	25.64
1971	12.90	0.84	19.84	21.97	11.81	39.10	21.00
Average	**10.26**	**4.02**	**15.25**	**17.17**	**9.81**	**27.35**	**19.61**
1972	13.32	3.00	19.71	20.79	11.43	45.05	36.33
1973	12.83	3.11	16.15	17.38	12.69	50.03	50.84
1974	1.16	2.19	–1.65	–5.40	3.24	25.79	83.70
1975	4.93	–4.09	6.24	4.35	6.01	–5.85	–14.56
1976	13.86	8.26	21.60	22.47	8.94	53.81	27.67
1977	10.19	3.79	12.41	12.88	9.59	14.63	12.00
1978	13.59	–1.04	18.36	20.65	12.21	35.53	29.56
1979	8.17	4.64	7.51	7.86	9.43	26.93	33.98
1980	7.30	–2.00	9.28	9.87	7.01	23.03	33.57
1981	6.16	–0.38	6.30	7.61	7.04	14.13	7.43
Average	**9.15**	**1.75**	**11.59**	**11.85**	**8.76**	**28.31**	**30.05**
1982	3.55	2.45	1.30	2.15	5.85	–1.80	–10.91
1983	8.45	1.78	9.80	11.50	8.15	13.15	7.41
1984	10.60	1.80	12.69	14.51	9.82	21.23	8.24
1985	4.95	2.22	3.66	3.30	6.51	0.89	–8.46
1986	11.64	–0.04	13.30	15.16	11.48	29.73	20.29
1987	12.74	5.86	12.43	13.08	13.75	34.66	44.67
1988	7.84	1.05	5.18	4.23	10.97	13.02	41.99
1989	8.23	–0.55	4.50	3.69	12.28	9.29	5.22
1990	5.39	2.08	1.09	–0.44	9.17	1.37	4.69
1991	7.55	1.76	6.82	6.71	8.54	13.34	14.89
Average	**8.10**	**1.84**	**7.08**	**7.39**	**9.65**	**13.49**	**12.80**
1992	6.76	–2.87	4.28	3.24	9.27	6.95	14.55
1993	6.32	5.44	4.12	2.51	7.93	4.44	7.02
1994	6.54	–4.37	5.69	5.73	7.77	9.35	10.76
1995	6.03	2.40	5.80	5.99	6.40	20.00	21.33
1996	5.70	0.04	3.90	4.50	7.20	3.90	–2.20

Source: *National Income of the Republic of China*, Directorate-General of Budget, Accounting and Statistics, Executive Yuan, ROC.

Table 4.2 Structural change in Taiwan (%)

Year	*Agricultural production*	*Industrial production*	*Manufacturing production*	*Service production*
1961	24.95	23.70	15.73	51.35
1971	13.84	36.84	28.76	49.32
1981	6.87	45.38	35.92	47.75
1986	5.12	45.91	38.30	48.98
1987	4.80	45.78	38.42	49.41
1988	4.50	44.65	37.13	50.85
1989	4.14	43.11	35.57	52.75
1990	4.01	41.35	33.60	54.64
1991	3.79	41.07	33.34	55.14
1992	3.45	40.11	32.24	56.44
1993	3.42	39.29	31.08	57.29
1994	3.07	38.97	30.85	57.96
1995	3.50	36.30	28.20	60.20
1996	3.28	35.64	28.04	61.08

Source: *National Income of the Republic of China*, Directorate-General of Budget, Accounting and Statistics, Executive Yuan, ROC.

time. However, with the sharp appreciation of the New Taiwan dollar in the second half of the 1980s, industrial production (mostly manufacturing) started to decline sharply, with its share of GDP also decreasing significantly to 35.6 per cent in 1996, with the services share rising again to 61 per cent. With the rapid industrialisation of Taiwan's economy, agricultural production declined, to less than 4 per cent in the 1990s.

The post-war industrialisation of Taiwan's economy was based on the solid foundation of the agricultural sector which had grown during Japanese rule from 1889 to 1949. Within the first few years of the 1950s, the Nationalist (KMT) government, which had recently retreated from mainland China, began reconstructing the war-torn economy of the newly recovered island. At that time, the export of agricultural products and US aid were the main sources of foreign exchange needed for importing materials and mechanical equipment from abroad. In order to avoid importing too much, the government adopted many typical import substitution policies, such as stringent import control measures, high tariffs and multiple exchange rates.

Furthermore, following the traditional infant-industry argument, import substitution policies were adopted to keep foreign products away from the domestic market to nurture Taiwanese enterprises in the protected domestic market. However, the import substitution policies did not work in the way that the government wanted them to. The economy did not grow very much because the size of the Taiwan market was too small to be able to nurture many local industries. The failure of the import substitution policies in developing the economy quickly enough led the government to realise that local enterprises had to look to the much larger market outside to be able to grow sufficiently large to take advantage of economies of scale. Subsequently, the government emphasised policies that encouraged and enabled domestic companies to sell their products in very competitive world markets.

In order to speed up economic development in Taiwan, the government implemented several policies that were mostly export-oriented. In fact, at the time, import substitution policies continued as export promotion policies were implemented, taking the leading role in the economic development of Taiwan (Wang and Wu *et al.*, 1988). First of all, trade restriction measures were largely removed and import tariffs were scheduled to be lowered gradually. The opening of the domestic market allowed local enterprises to learn from competing imported products. Very soon, the local people were able to produce the same products that could be sold in the world market at very competitive prices. Second, a variety of policies and measures were designed and implemented to promote exporting industries. The overvalued and multiple-rated currency system was simplified and depreciated against the US dollar to around 36–40 to 1 from 1960 until 1985 when the New Taiwan dollar began to appreciate sharply to 26 to 1 in 1988. Imported machinery, materials and semi-finished products were exempted from taxes if they were used for producing exported goods, greatly lowering the costs of production. A tax refund scheme – which covered the defence tax, harbour dues and excise tax – was implemented to help reduce the costs of exporting goods. According to Wang and Wu and their colleagues (1988), these tax refunds amounted to 62 per cent of all tax revenue in 1972.

Furthermore, in order to help local enterprises finance their exporting business, low-interest loans were provided for exporters by the banks. These loans have been very popular and very critical for the presence of numerous small and medium-sized enterprises (SMEs) in Taiwan. At that time, there were no privately-owned banks; all the banks were government owned and willing to comply with the export-promoting policy. Exporters could easily get a loan from the banks once they received a letter-of-credit issued by the foreign buyer and shipping documents. These low-interest loans played a critical role in helping the mostly small and medium enterprises finance their business.

Third, three export-processing zones (EPZs) were constructed to attract multinational corporations to move their production bases to Taiwan. For foreign firms located in the EPZs, import and export procedures were kept as simple as possible to reduce the administrative costs of producing in Taiwan. No duties were levied on imported machinery and goods for reprocessing in the zone. The first such export-processing zone was established in 1966 in Kao-shung and two more were added in 1971. Foreign capital poured into Taiwan to take advantage of the ample supply and relatively inexpensive labour and the favourable tax incentives provided by the government. These foreign investments not only created a lot of jobs and earned much needed foreign exchange for Taiwan, but also brought much needed technology into Taiwan. These technologies and well-trained engineers and workers quickly spread to other parts of Taiwan as people moved around or started their own businesses (Ranis and Schive, 1985; Wu, 1989). Since big multinational electronic firms were the largest investor at the time, the transferred technologies gave Taiwan a healthy start when the government decided to develop the information technology industry.

The export-promotion policies helped Taiwan's economy to take off rapidly in the 1960s and 1970s (see Table 4.1). The annual growth rate of exports averaged 27.4 per cent and 28.3 per cent respectively. The expansion of exports not only helped the economy accumulate much needed foreign exchange but also raised Taiwan's economic growth rate to 10 per cent over the two decades. In reviewing this development process, it becomes clear that implementation of export-promotion policies was the most effective way of developing an economy with a small domestic market. The export-oriented policies exposed domestic enterprises to the competitive world market so that they had to learn how to produce goods as cheaply as possible in order to survive. In the end, market forces led local enterprises to concentrate in those sectors in which Taiwan had a comparative advantage at that time including products such as textiles, clothing, shoes, toys and consumer electronics, most of which are highly labour intensive.

The fast expansion of exporting industries created a huge demand for raw materials, intermediate goods and machinery used for producing the goods. The need to fulfil this demand domestically brought about the so-called second-round of import substitution in Taiwan (Schive, 1990). With the wealth and experience accumulated during the take-off period led by the fast growth of exports, local enterprises were ready to enter into more technology and capital intensive sectors such as machinery, petrochemicals and artificial fibre industries. As can be seen from Table 4.3, these industries have played an important role in Taiwan since the 1970s.

During this second-stage of import substitution, the policies adopted by the government were quite different. In order to enhance the competitiveness of domestic enterprises, the government decided to invest in several very large projects to improve the infrastructure and to facilitate the development of technology and capital intensive industries in Taiwan. These projects included a highway system across the island from north to south, a new sea port in the middle part of Taiwan and a new international airport. Other projects that were more directly related to industry included an integrated steel mill, a shipbuilding yard and a new refinery to help expand the capacity of the petrochemical industry. All these projects were implemented to either increase efficiency or to reduce the production costs of local enterprises. Starting from the mid-1970s, the government again took the leading role in a series of research and development projects in the field of integrated circuits (ICs). As we will discuss in more detail, these pioneering projects to develop IC technology greatly helped lay the foundation for Taiwan to become an important IT player.

In the mid-1980s, a couple of critical events transformed the economy once again. First, the New Taiwan dollar appreciated sharply within a short period, with the exchange rate with respect to the US dollar going up from 40 to 1 in 1985 to 36 to 1 by the end of 1986 and 29 to 1 in 1987. The NT$'s swift appreciation greatly reduced the competitiveness of Taiwanese products in the world market. Second, a labour shortage and the increasing cost of other production factors, such as land

Table 4.3 Structural change of the manufacturing sector in Taiwan (%)

	1951	*1961*	*1971*	*1981*	*1986*	*1987*	*1988*	*1989*	*1990*	*1991*	*1992*	*1993*	*1994*	*1995*
Manufacturing	**100.00**	**100.00**	**100.00**	**100.00**	**100.00**	**100.00**	**100.00**	**100.00**	**100.00**	**100.00**	**100.00**	**100.00**	**100.00**	**100.00**
Food	29.34	25.22	11.35	9.93	9.22	9.12	8.79	8.33	8.57	7.71	7.97	7.54	7.48	8.35
Beverage and tobacco	12.32	13.54	7.71	2.83	1.83	1.90	1.65	1.71	1.67	1.59	1.65	1.60	1.57	0.49
Textile	15.16	11.13	11.83	9.43	8.44	8.24	7.12	6.90	6.70	6.83	7.10	6.63	6.31	6.29
Wearing apparel	1.86	2.85	6.08	6.12	5.94	5.13	4.34	4.15	4.15	3.43	3.10	3.24	2.44	0.90
Leather	0.60	0.28	0.60	1.29	1.90	1.64	1.61	1.42	1.29	1.30	1.06	1.04	0.92	0.83
Lumber and furniture	3.56	5.19	3.97	3.30	3.19	3.35	3.11	3.02	2.67	2.58	2.26	1.98	1.68	0.52
Paper and printing	6.51	6.81	4.30	4.51	4.04	3.93	4.07	4.25	4.00	3.76	3.66	3.42	3.46	0.87
Chemical*	16.64	13.01	20.62	13.87	15.18	15.67	16.12	14.74	13.04	16.45	15.18	15.20	16.85	17.00
Rubber products	1.48	1.01	1.12	1.37	1.29	1.26	1.28	1.29	1.38	1.32	1.39	1.39	1.23	1.07
Plastic products	0.00	0.00	0.00	5.44	7.01	6.90	6.61	6.70	6.53	6.26	6.08	6.01	5.76	5.47
Non-metallic mineral products	3.94	7.27	4.46	4.81	3.42	3.45	3.78	4.23	4.42	4.24	4.70	5.24	5.06	3.35
Basic metal	1.31	4.45	3.40	5.34	5.74	5.62	6.58	7.06	6.86	6.51	6.47	7.20	7.17	8.73
Metal products	0.27	1.47	2.05	3.68	4.35	4.48	4.78	5.22	5.94	6.20	6.73	6.85	6.96	5.38
Machinery	1.31	2.13	3.36	3.03	3.16	3.47	3.90	3.86	4.46	4.50	5.14	5.12	5.09	4.40
Electrical machinery and appliances	0.55	1.91	10.14	11.44	13.03	13.57	14.07	14.75	15.93	15.64	15.45	16.34	17.04	22.63
Transportation equipment	1.26	2.48	4.91	6.88	5.56	5.98	6.30	7.10	7.50	7.40	8.08	7.58	7.80	7.72
Precision instrument and miscellaneous	3.89	1.25	4.09	6.73	6.71	6.26	5.88	5.25	4.89	4.26	3.95	3.62	3.18	2.42

Source: *National Income of the Republic of China*, Directorate-General of Budget, Accounting and Statistics, Executive Yuan, ROC.
Note: * Including chemical materials, chemical products and petroleum and coal products.

and labour, severely exacerbated the comparative disadvantage of Taiwan's exports. In order to survive the impact of these two events, Taiwanese enterprises started to search abroad for less expensive labour and land for production. An exodus has occurred in the past ten years, with a lion's share of investment going to the Southeast Asian economies and to mainland China.

Since the 1960s, manufacturing has become the main pillar of Taiwan's economy. With rapid development of the economy, the industrial structure of Taiwan has changed significantly over time (Table 4.3). The share of labour-intensive industries (including food, beverage and tobacco, textile and garments, leather, lumber and furniture, paper and printing) has declined sharply from 65 per cent in 1961 to nearly 24 per cent in 1994.[2] The production of capital-intensive and technology-intensive industries – such as chemicals, basic metal products, machinery, electrical machinery and appliances – has increased at a much greater pace during this period. Among capital-intensive industries, electrical machinery and appliances, and more specifically computer-related information industries, have become the driving force of Taiwan's economic growth in the 1990s. Production of electrical machinery and appliances accounted for 22.6 per cent of GDP and 36.2 per cent of exports in 1995.

A few more remarks about Taiwan's economy are in order. First, Taiwan is a small country with limited arable land and very high population density. The geographical area is about 36,000 square kilometres, and the population is more than 20 million. With the size of Taiwan's domestic market, looking outwards seems to be the only solution for domestic enterprises. The government, led by some very talented bureaucrats in charge of policy in the early stage of economic development, was well aware of the importance of nurturing the competitiveness of domestic enterprises for the world market and designed development policy accordingly.

Second, as the government put more emphasis on export-oriented policies, local enterprises were encouraged to invest in those industries in which Taiwan had comparative advantage and to produce in a labour-intensive way. In the 1960s and early 1970s, with a rather abundant and well-educated labour force, Taiwan had a very strong comparative advantage in labour-intensive industries such as garments, shoes, sporting goods, sportswear, bamboo products and low-end textiles. But, when labour costs began to rise in the 1970s, it became too costly to produce goods that required a lot of labour. The changing comparative advantage in Taiwan required the economy to produce more capital and technology-intensive, instead of labour-intensive products. As shown in Table 4.4, textiles and garments were Taiwan's largest export items. Together, they accounted for more than one-third of total exports in 1971, while electronic machinery and machinery products made up only 16 per cent. Plywood contributed almost 5 per cent of Taiwan's exports at the time. By 1981, footwear, toys and sporting goods had become important exports with electronic products accounting for the largest share of exports. Information and communications products only became significant later, contributing 7.3 per cent of Taiwan's total exports by 1991, with its importance increasing in the 1990s.

Table 4.4 Taiwan's top five export products (%)

Rank	*1971*	*1981*	*1991*	*1996*
1	Textiles and garments (35.4)	Electronics (13.0)	Textiles (11.1)	Electronics (14.3)
2	Electrical machinery (12.9)	Garments (11.3)	Electronics (10.7)	Textiles (11.5)
3	Food (4.7)	Textiles (9.8)	Machinery (8.9)	Information and communications (10.8)
4	Plywoods (4.6)	Footwear (6.8)	Information and communications (7.3)	Basic metals (8.8)
5	Machinery (3.2)	Toys, games and sports (5.9)	Basic metals (6.3)	Machinery (8.2)

Source: GCEPD, *Taiwan Statistical Data Book, 1996*, Taipeh.

Third, events since the second half of the 1980s had made it necessary for the economy to move quickly in response to pressures for liberalisation and globalisation. For example, most services were regulated or even monopolised by the government. New private banks were allowed to operate from 1991; oil refining, electricity and telecommunications, that used to be only operated by government-owned public enterprises, were opened to the private sector, and even to foreign capital. Internationally, Taiwan became a major investor in Southeast Asia and mainland China. Accumulated investment in the Asian region (excluding mainland China) from 1952 to 1986 was only US$87 million, but increased to US$4.7 billion by the mid-1990s, with three-quarters going to Southeast Asia. However, the biggest increase in outward investment by Taiwanese enterprises has been in mainland China, with total investment amounting to US$6.8 billion since the government allowed indirect investments in mainland China.[3]

Some 'star industries' have emerged at different moments in Taiwan's economic development. These industries were usually very competitive and soon gained large shares in international markets. In the early days, clothing, shoes, umbrellas, plastic goods, sporting goods, black and white TVs, video games, calculators and bicycles were significant; electronics, PCs and computer peripherals have emerged in more recent years. The sequence of emergence of these products as important export items at different stages of economic development reflect changing comparative advantage, partly due to changing policy guidelines for the private sector. Taiwan started industrialisation with labour-intensive industries that used simple, easy-to-learn production technologies, when labour was still relatively abundant, capital was scarce, and industry's technological level was quite modest. Gradually, with the accumulation of capital and the rising technology level, new industries and product lines involving higher technology and greater capital intensity were used.

IT industry development

Taiwan's information technology (IT) industry demonstrates how internationally competitive its manufacturing enterprise has become in the past two decades.[4] Estimated total production of the industry reached around US$19 billion in 1997,[5] up from US$2.1 billion in 1986. Its share of GNP also increased from 2.8 per cent to 6.6 per cent in this period. In terms of exports, the IT industry contributed 5.2 per cent to national exports in 1986, while the ratio increased to around 20 per cent in 1997, making IT the leading export industry in Taiwan. Several products in the IT industry, such as portable computers, monitors, motherboards, keyboards, scanners, mice, etc., ranked number one in the world in terms of output (Table 4.5). The following brief review of the recent development of (both personal and portable) computers and their peripherals shows how Taiwan has achieved competitiveness in the world market.

Computer-related information products

Computers and related products have become the most important industry in Taiwan in recent years. Total domestic production of these industries in 1995 was US$14.2 billion, 22 per cent up from 1994 and accounting for 5.4 per cent of GNP. This made Taiwan the third largest producer of IT products in the world, only surpassed by the United States and Japan (Table 4.6). These industries accounted for about 12 per cent of total Taiwanese exports in 1995, up from 5.2 per cent in 1986.

In 1996, Taiwanese companies (including overseas subsidiaries) produced 5.87 million desktop computer sets (a 28 per cent increase from 1995) and 3.77 million sets of portable computers (a 45 per cent increase from 1995). Taiwan accounted for 10 per cent of the world market for desktop computers, and ranked third in the world market. For portable computers, Taiwan accounted for 32 per cent of the world market, up from 27 per cent in 1995, and surpassing Japan to become the number one producer in the world.[6] Almost 80 per cent of the portable computers were produced under original equipment manufacturing (OEM) or original design manufacturing (ODM) contracts. Both American firms (such as Compaq and IBM) and Japanese firms (such as NEC, Hitachi and Epson) are major contractors of Taiwanese producers. Half of the computers produced in Taiwan were exported to the United States, while 30 per cent went to Europe.

Only a few Taiwanese computer manufacturers are big companies (with capital of more than NT$60 million or employing more than 200 workers), with the majority consisting of small and medium enterprises. In 1995, the top ten computer manufacturers produced more than 80 per cent of Taiwanese-made computers. The biggest computer producer in Taiwan is Acer, which produced 1.6 million computers for other world-known computer companies and sold 1.19 million under its own brand name in 1995. With total sales of 1.81 million of its own-brand personal computers in 1996, mainly to Third World countries, Acer was the eighth largest computer supplier in the world in (Table 4.7).[7]

Table 4.5 Status of Taiwan's top-ranking computer products, 1996 (000 units)

Products	*Share of Science-based Industrial Park (1994)*	*Domestic output*	*Overseas output*	*Total*	*World market share (%)*	*Ranking*
Monitor		17,735	17,108	34,843	53.4	1
Portable computer	31.0	3,772	0	3,772	32.0	1
Desktop computer	21.10	5,025	845	5,870	10.2	3
Motherboard	29.39	18,823	12,497	31,320	74.2	1
Image scanner	73.43	3,750	0	3,750	61.2	1
Graph card	—	6,900	4,600	11,500	38.3	—
LAN card	63.42	9,946	318	10,264	38	—
Terminal	66.12	800	0	800	24.6	—
Switching power supply	—	6,662	31,848	38,510	55.3	—
CD-ROM Optical disc driver	—	4,500	2,100	6,600	15	—
Video card	—	544	256	800	55	—
Mouse	39.84	14,430	30,663	45,093	65	1
LAN Hub*	100.00	922	11	933	22	—
Sound card	—	8,360	1,140	9,500	50	—
Keyboard	—	2,964	39,374	42,338	61	1

Source: Market Intelligence Center, Institute for Information Industry.
Note: * For 1995.

Table 4.6 Major information technology (IT) producers, 1994–5 (US$ million)

	1994	*1995*	*Growth rate (%)*
USA	60,307	65,132	8
Japan	61,107	63,551	4
Taiwan	11,579	14,156	22
	(3,003)	(5,511)	(84)
Singapore	11,420	13,480	18
United Kingdom	8,952	9,668	8
Germany	7,135	7,492	5
France	6,560	6,822	4
Italy	6,394	6,650	4
Brazil	5,837	6,084	4
Korea	4,320	5,079	18

Source: GCEPD, *Taiwan Statistical Data Book, 1996*, Taipeh.
Note: The number in parentheses is for the overseas output of Taiwanese firms.

Table 4.7 World's top ten PC suppliers, 1996

Rank	*Company*	*Sales (million sets)*	*Market share (%)*
1	Compaq	5.73	9.9
2	IBM	4.80	8.2
3	Apple	4.63	8.0
4	NEC	3.04	5.2
5	Packard Bell	2.99	5.1
6	HP	2.04	3.5
7	Dell	1.81	3.1
8	Acer	1.81	3.1
9	Toshiba	1.47	2.5
10	Fujitsu	1.43	2.5

Source: GCEPD, *Taiwan Statistical Data Book, 1996*, Taipeh.

Monitors

The technology for producing monitors has become so mature and common that it is no longer a severe entry barrier. Instead, economies of scale have become very critical in the industry. With years of experience manufacturing large-volume-production electronic goods such as televisions, calculators and telephone sets, Taiwanese firms have gained strong competitive advantage in the monitor industry. The total production of Taiwan's monitor industry (including production in overseas subsidiaries) more than tripled from 1991 to 1996. In 1991, Taiwanese firms shipped out 9.8 million sets of monitors, accounting for 39.8 per cent of total world shipments. In 1996, shipment of monitors from Taiwanese firms increased to 34.8 million sets, i.e. 53.4 per cent of world shipments.

Among the top ten monitor producers in the world, seven are in Taiwan. The largest monitor producer in Taiwan is Philips, a foreign-owned subsidiary of Philips in Holland. It produced 4.7 million sets in 1995 while the largest domestically-owned firm had a production capacity of 3.5 million sets. All the top five domestically-owned monitor producers in Taiwan have annual capacities of more than 2 million sets, while the top ten producers now account for more than 90 per cent of monitors produced in Taiwan. In order to enhance their competitiveness, these large producers all have expansion plans for the next few years, either domestically or overseas.

Like most of the other IT industries, overseas production has become more important for Taiwanese monitor producers because of the narrow profit margins in the industry and the cost disadvantage compared to low-cost overseas production. In order to remain competitive, Taiwanese enterprises have had to move production of small monitors abroad. In 1996, overseas production accounted for almost 50 per cent of total Taiwanese firm output. Most of these overseas plants are located in Southeast Asian countries or mainland China, while a couple of new plants have been set up in Scotland and Mexico.

Other IT products

Motherboards were the fourth largest item among the top-ranking IT products manufactured in Taiwan (Table 4.8). Total production of motherboards in 1997 was worth US$3.8 billion, up from US$2.2 billion in 1995, i.e. a 70 per cent increase. In 1996, 74 per cent of world motherboard production was controlled by Taiwanese firms, up from 65 per cent in 1995. Of these, 13 million units were

Table 4.8 Output growth of Taiwan's leading computer products, 1995–7 (US$ million)

Product	*1995*	*1996*	*1997*	*Growth rate (1995/6)*	*Growth rate (1996/7)*
Monitor	7,271	7,872	7,919	8.27	0.60
Portable computer	3,339	5,331	6,846	59.66	28.40
Desktop computer	2,314	3,372	5,125	45.72	52.00
Motherboard	2,222	3,075	3,784	38.38	23.00
Switching power supply	895	1,078	1,297	20.45	20.30
Terminal	712	855	1,050	20.08	22.80
CD-ROM	305	528	1,023	73.11	93.60
Image scanner	536	786	922	46.62	17.30
Keyboard	369	415	497	12.47	19.76
Graphics card	516	558	499	8.14	–10.60
LAN Hub	203	265	360	30.54	35.80
Sound card	123	228	167	85.37	–27.00
Mouse	191	198	174	3.66	–12.12
Video card	199	100	92	–49.75	–8.00
Total	19,583	25,065	30,157	27.99	20.30

Source: Market Intelligence Center, Institute for Information Industry.

produced locally and 7.7 million units were produced in overseas subsidiaries, i.e. the ratio of overseas production was 39.8 per cent in 1996.

CD-ROMs registered the highest growth rate in the mid-1990s. In 1995 and 1996, production almost tripled from US$305 million to US$1,023 million. Desktop computers and portable computers have also had rather high growth rates in 1995 and 1996. Production of both items more than doubled from 1995 to 1997. However, some items, such as mice, video cards, sound cards and graphics cards saw negative growth rates in 1997.

Wade (1990) has observed that the Taiwan government exercised substantial leadership in promoting industrial growth. Schive (1990) has shown how the government helped lay the foundation for the development of the IC industry in Taiwan. Before exploring the reasons why the information technology industry developed so successfully in such a short period, the following two sections will introduce some important institutional infrastructure created by the government and their role in the development of the information industry.

Government industrial technology promotion efforts in Taiwan

In order to transform Taiwan more swiftly into an industrialised economy, the government not only put great efforts into designing and implementing various macroeconomic policies, but also played an important role in the advancement of industrial technology. A variety of technology-upgrading policies were carried out by different institutions. Of these, the National Science Council (NSC) under the Executive Yuan, as well as the Industrial Development Bureau (IDB) and the Department of Industrial Technology (DIT), both of the Ministry of Economic Affairs (MOEA), have been the most important agencies for the advancement of science and technology in Taiwan. The NSC is responsible for setting the goals and priorities for technology development in Taiwan, and the two agencies within the MOEA for implementing the plan.

The Industrial Technology Research Institute (ITRI) and the Institute for Information Industry (III) are two institutes responsible for the development of hardware and software information technology respectively, while the Hsin-chu Science-based Industrial Park (HSIP) has been the cradle of Taiwan's information technology industry since its establishment in 1980. The HSIP was founded and operated by a special administration office, under the NSC.

Industrial Technology Research Institute (ITRI)

ITRI was founded in 1973 as a non-profit research organisation under the supervision of the MOEA. It is located in the Hsin-chu area, about 80 kilometres from Taipei. The mission of the institute is to attend to the private sector needs by undertaking research and development (R&D) of key technologies for them. ITRI has become a technical centre for the industry, is now staffed by more than 5,000 research personnel and has an annual budget of around US$500 million.

Since its inception, ITRI has become a very important instrument for advancing technological capability in Taiwan. ITRI has grown into an institution with ten divisions covering the following important industrial fields: electronics, optic-electronics, consumer electronics, computers, ICs, communications, materials, energy and resources, aerospace, machinery, and chemicals. ITRI activities range from technology development and prototype manufacturing, to the provision of technical assistance and consultation for the private sector. Its electronics-related institutes – such as electronics, optic-electronics, computer, communications and consumer electronics – have taken up one-third of total government funding. Annually, numerous projects on the technological development of various industries have been conducted by ITRI, either alone or with participation by the private sector. Technological innovation is transferred to the private sector to enhance its competitiveness. The development of the IC industry in Taiwan illustrates the key role that ITRI has played during the past two decades.

By the early 1970s, when it became clear that labour-intensive industry would no longer be able to carry Taiwan's economy much further, the government began to contemplate what kind of industry would lead economic development next. After long deliberation, the government decided to turn to electronics-related industries. The development of technology for designing and manufacturing integrated circuits (ICs) was chosen as a pioneer project. In 1975, the Electronics Research and Service Organization (ERSO) of ITRI was commissioned with the task of developing advanced IC technology for Taiwan, which was considered to be a 'mission impossible' at that time. In 1976, after long consultation with several 'overseas' Chinese experts in the field, ITRI decided to sign a technical transfer contract with RCA, then a leading firm in the world IC industry. In this contract, RCA agreed to transfer IC technology, including design, processing, testing and production management. In order to facilitate the successful transfer of technology, RCA also provided quite comprehensive training programmes for engineers from Taiwan. With the technical assistance of RCA and the hard work of a group of young, talented and hard-working engineers, ERSO was able to accomplish the task within a period of four years.

Once the development of IC technology reached the commercialisation stage, the government decided to establish a company to commercialise the newly-developed technology. Since the life cycle of IC products is usually very short and the technology evolves rather rapidly, it would be wrong for ITRI to hold on to the technology itself rather than let a private company take over the technology and develop it into marketable products. Therefore, a company was founded in 1980 named the United Microelectronics Company (UMC).

Operating as a spin-off, the UMC not only received all the facilities and technologies, but also most of the people involved in the pioneering project of developing the first IC technology in Taiwan. UMC was founded with US$10 million and most of the capital was from the government or semi-government institutions. At that time, the private sector had neither much interest nor confidence in the IC industry. Therefore, only a small portion of the capital was from local consumer electronics producers, whereas the majority of the capital came from the MOEA, the Chiao-Tong Bank (a government-owned investment bank), the China Development Corporation (CDC) and Kung-Hwa Securities, both semi-government agencies.

In the early days of UMC, ITRI not only transferred all the technologies it developed, but also worked closely with UMC to provide technical services, construction supports, manpower training and production management to the new-born company to make sure that the technology would take root and could be further developed in UMC. UMC started with technology to design and manufacture ICs for consumer electronic goods such as calculators, electronic watches, video games, telephones, etc. These products were the main export items of Taiwan's electronic industry at the time. The surging demand in these domestic industries provided very important support for a newly established domestic company like UMC, because it allowed UMC to gain more experience before it had to compete with other foreign firms in the world market. UMC has become much more experienced and bigger, employing over 2,750 people with capital reaching US$12 billion.

Total UMC revenue was less than US$5 million in 1982, and reached US$0.9 billion in 1995, i.e. with average annual revenue growth from 1982 to 1995 surpassing 40 per cent. In the third year, UMC began to make profits of US$25 million from its operations. Since then, UMC has been on a fast growth track to become one of the leading IC firms in Taiwan. Over the years, UMC has developed strong capabilities in designing and manufacturing a wide range of sophisticated and powerful ICs. In addition to the foundry service, UMC product lines include a wide range of memory and commercial ICs. UMC is one of the largest IC producers in Taiwan, second only to TSMC – another ITRI spin-off.

The successful experience of UMC encouraged the government to proceed with two more R&D projects on more advanced IC technology. The first project gave birth to the world's largest foundry firm, the Taiwan Semiconductor Manufacturing Company (TSMC), and the Vanguard International Semiconductor Corporation (VISC). Following the footsteps of UMC, a number of smaller IC firms also came onto the market in Taiwan. These firms mainly focused on providing testing, packaging, mask-making and designing services for UMC and other foreign firms. Together with UMC, TSMC and VISC, these firms helped to form the backbone of Taiwan's fast-growing IC industry.

By the second half of the 1980s, a global market for a professional semiconductor foundry was emerging. In order to meet the demand for such a service, the government founded the TSMC in 1987, to serve as the first pure IC foundry company in the world. The company is a second spin-off from ITRI after a five-year research project on very large scale integrated circuit (VLSIC) technology was completed. The total budget for this project was more than US$100 million. This time, about 150 persons transferred directly from ITRI to the new company. The facilities and equipment used for developing the technology were provided to the TSMC through a rental agreement. TSMC now has more than 5,400 employees, half of them holding college or higher degrees.

Morris Chang, a veteran of the IC industry in the United States, was recruited to serve as chairman of the TSMC board of directors. His expertise has greatly helped in achieving the goals set for the TSMC. By 1994, TSMC had already captured nearly 20 per cent of the world's IC foundry service market, with customers including big company names such as Motorola and Intel. By 1997, it had a monthly production capacity for manufacturing over 110,000 pieces of 6-inch and 8-inch wafers, and was about to add a further 60,000 piece capacity to its production line soon. This maintained TSMC as the number one foundry company in the world. Total sales of TSMC were US$1.36 billion in 1996, surpassing UMC to rank number one in Taiwan's IC industry in 1995.

The Vanguard International Semiconductor Corporation (VISC) is a joint venture between the MOEA and other domestic enterprises, including TSMC. The scale of this government-initiated research project is much bigger than its two predecessors. With a total budget of US$256 million, the purpose of the project is to develop technology for sub-micron IC manufacturing of 8-inch wafers. Once completed, the technology will make Taiwan a front runner in the very dynamic IC industry. VISC was founded in 1994 to advance Taiwan's IC industry into the then leading edge 0.25um and 0.18um technology, with an annual manufacturing capacity of 1 million pieces of 8-inch wafers. A total of 350 people, with various levels of training and skills, moved to this new company.

Since the establishment of ITRI in 1973, the government has spent more than US$5 billion to finance research projects. About a third of these funds has been allocated to electronics-related projects. The three IC projects introduced here are some of the largest projects ever carried out by ITRI. Usually, ITRI develops technologies that the government believes to be instrumental for enhancing the competitiveness of the private sector, and not many such technologies are transferred through spin-off companies, i.e. the three IC technology spin-off cases are not typical of ITRI's style of operation. Although ITRI's performance has been subject to constant criticism (Wu *et al.*, 1986), ITRI has played a critical role in the development of Taiwan's IC industry. The three ITRI spin-offs have successfully nurtured the development of Taiwan's IC industry. After the fast growth of UMC and TSMC, many domestic enterprises participated in the downstream activities of IC packaging and testing, and the upstream businesses of IC designing and photo-masking. There are even three firms producing wafers for the industry. UMC and TSMC are two key players in the industry, and both are well known for their strong capabilities in providing services for leading international IC firms like Motorola and Intel. Taiwan is ranked number four (after the US, Japan and South Korea) in the world in terms of IC production, and more importantly, the industry has been very competitive and is able to support the development of PC-related industries in Taiwan. Both UMC and TSMC had several big investment plans, which would, when completed, expand their capacities by at least four times from production levels in 1996.

Institute for Information Industry (III)

The IT industry can be divided into manufacturing of hardware on the one hand and the computer systems and applications on the other. Since the development of the two are highly related, in addition to the establishment of ITRI to help develop hardware technology for the private sector, the government also established the Institute for Information Industry (III) in 1979 to promote computer usage and to build software design capability in systems programming. To advance the use of information technology in both government agencies and the private sector, III does strategic planning for the information industry, providing assistance for government and private enterprise computerisation, promoting information technology applications, providing training services for both public and private sectors, and providing market intelligence services.

The Institute of Information Industry is now staffed by more than 1,200 professionals. Half its employees hold a Masters or higher degree and over three-quarters have IT-related backgrounds. Of the III's functions, market intelligence services may be the most critical. The Market Intelligence Center (MIC) of the III is the principal agency responsible for gathering and analysing market information for information-related industries in Taiwan. Reports on the world markets for information industries are issued regularly, and seminars on new technologies are also held regularly. Also, through co-operation with leading international software companies such as IBM, Hewlett Packard, Unix System Labs, AT&T, etc., III serves as a conduit for transferring leading-edge computer software technology to Taiwan (Kraemer *et al.*, 1996: 231).

Hsin-chu Science-based Industrial Park (HSIP)

The establishment of the Hsin-chu Science-based Industrial Park (HSIP) was another critical institutional initiative for advancing the development of the IT industry in Taiwan. The HSIP was formed in 1980 with the intention to create a 'Silicon Valley' in Taiwan.[8] By establishing ITRI, the government intended to enhance the competitiveness of local enterprises by helping them develop key technologies as well as by providing them with the necessary technical services. The mission of the HSIP has been to facilitate an attractive and conducive environment for the high-tech industries identified by the government to be established and promoted in Taiwan. The high-tech industries designated by the government include communications, information, consumer electronics, semiconductors, advanced and precision machinery, aerospace, advanced material, special chemicals, medical instruments and environmental industries.

An administrative office, with the full authority of the National Science Council,[9] has been put in charge of providing all kinds of services for incoming investments. An advisory committee has been formed to supervise HSIP operations, and all investment projects have to be approved by the committee. Only investments in designated high-tech industries can be located in the Park. Over the past fifteen years, the government has allocated a total of US$483 million to the Park. The administrative office is designed to process all the applications needed for a new investment. Once the investment is approved by the advisory committee, the office will assign land to the firm, assist in plant construction and even train the workers. The land is provided on a rental basis to lower production costs.

The Park office also provides services to assist Park-based enterprises (PBEs) in finding engineers and skilled workers as well as on-the-job training for their employees. In fact, there are plenty of training courses for varied fields all the year round. These services have been very helpful in upgrading the productivity of workers and, more importantly, new technologies, in both production and management, are easily disseminated in the Park area. In order to facilitate the operation of the PBEs, several business support service agencies are also available in the Park. There is a special customs office in the Park for administering the import and export business for PBEs, so that they can clear their goods right in the Park. Banks are also available in the Park to provide financial services, as are a post office and a telecommunications station to help with communications. In addition to these business-related services, the Park also provides special educational facilities for the children of returning expatriates. An experimental educational system from kindergarten to senior high school was founded in 1983. Bilingual programmes are provided, especially to help expatriate children adjust to the new learning environment.

By establishing the HSIP, the government sought to transform Taiwan into a 'science and technology island'. The HSIP is located in the Hsin-chu area in the vicinity of ITRI and two outstanding universities. The objective of creating an industrial park such as HSIP was to induce investment, either domestic or foreign, in high-tech industry so that Taiwan's economy could advance further into higher value-added production. In order to encourage such investments, the government also provides very generous tax incentives and financial support for the PBEs. A newly-established PBE is entitled to a five-year exemption from corporate income tax. If the company invests further, there is another four years' exemption for the amount of the additional investment. According to our interviews, the effective corporate tax rates for all PBEs were 1.09 per cent and 2.58 per cent for 1994 and 1995 respectively; the average effective corporate income tax between 1990 and 1994 was 4.07 per cent, compared to 15.29 per cent for the top 100 manufacturing firms in Taiwan. In addition to tax incentives, low-interest loans (two percentage points below regular interest rates) are available for PBEs. Various kinds of grants are available for research and development activity and for component and product innovation by the PBEs. Furthermore, no duties and commodity taxes are levied on imported machinery, raw materials, fuel, supplies or semi-finished products; while sales tax is exempted for exported goods.

The establishment of the HSIP has been a great success by most accounts. In the first year, only 17 firms were set up in the Park with paid-up capital of only US$18 million. By the end of 1995, there were 180 companies located in the Park, with an aggregate investment of US$5.4 billion (see Table 4.9). Of these companies, 36 were foreign-owned and 144 were domestic firms. More significantly, the private sector invested 80 per cent, while the government accounted for only 8 per cent, with the rest coming from abroad. In terms of production, the PBEs also registered extremely high growth rates compared to the rest of the economy. Total output in 1986 was US$450 million, representing 0.5 per cent of total manufacturing production in Taiwan. But by 1995, the value of production had increased

Table 4.9 HSIP: approved investments, 1981–95 (US$ million)

Year	*No. of companies*	*Registered capital*	*Paid-up capital*
1981	17	22	18
1982	26	38	28
1983	37	59	48
1984	44	110	81
1985	50	144	102
1986	59	177	151
1987	77	546	332
1988	94	769	561
1989	105	1,345	1,072
1990	121	2,149	1,590
1991	137	2,377	2,059
1992	140	2,952	2,503
1993	150	3,072	2,495
1994	165	4,403	3,527
1995	180	7,088	5,400

Source: HSIP, *Annual Report 1995*.

to US$10.9 billion and the ratio to more than 3 per cent. By 2003, the production value of PBEs is expected to reach US$50 billion and to account for 13 per cent of manufacturing output. In terms of imports and exports, the increased role of the PBEs was even greater. Between 1985 and 1995, their import share rose from 0.7 per cent to 6.4 per cent, while their export share rose from 1.0 per cent to 6.8 per cent.

However, most importantly, the Park has become the centre for Taiwan's information technology industry. The HSIP hosts all the IC manufacturing companies in Taiwan, with only a few testing and packaging companies scattered around the country. One-third of portable computers, one-fifth of desktop computers, 30 per cent of motherboards and over 70 per cent of colour scanners were produced in the Park by the mid-1990s. Since more than 90 per cent of the Park's products is for export, we would expect high export ratios for each individual industry.

By 1995 sixteen companies had been listed on the Taiwan stock exchange and more were expected to go public. The increasing number of PBEs going public suggests that these companies have now earned the confidence of investors. The injection of public funds into these newly established firms allowed them to expand more quickly and to become more competitive in the world market. One more important feature of the PBEs is that they spend more on R&D than other firms located outside the Park. The ratio of R&D expenditure to revenue for Park-based companies was 4.6 per cent in 1994, well above the less than 2 per cent average for the whole island (Table 4.10).[10]

Thus, the HSIP is the cradle of Taiwan's information technology industry. Three factors are of great importance to the HSIP's success. First, the Park provides a

Table 4.10 HSIP: R&D investments of Park-based enterprises, 1994

Industry	*Expenditures (US$ million)*	*Expenditures/ sales (%)*	*R&D personnel*	*Percentage of employees*
Integrated circuits (ICs)	174	5.5	1,746	10.7
Computers and peripherals	75	2.8	1,327	13.8
Telecommunications	37	6.6	547	13.9
Opto-electronics	16	9.3	235	9.9
Precision machinery and materials	5	6.5	106	10.6
Bio-technology	3	19.1	58	23.3
Total	310	4.6	4,019	12.0

Source: HSIP, *Annual Report 1995*.

unique environment for high-tech enterprises to start up and to grow. Following the successful experiences of export-processing zones (EPZs) established in the 1960s and 1970s, the government designed the Park as a special industrial zone to accommodate the needs of high-tech industry. The HSIP administration office provides all kinds of services from land and plant construction, to technical and management personnel training. Important services, such as banking, telecommunications, and customs, etc., are also available in the Park. The one-stop services provided in the Park have greatly increased the convenience of operating park-based enterprises.

Second, the Park is surrounded by two top-ranking universities, ITRI and four national laboratories. In fact, three of the laboratories are located within the Park. Two important research and supporting agencies are also located in the Park: the Precision Instrument Development Center and the Chip Implementation Center. National Chiao-Tong University and National Tsing-Hua University are two outstanding universities in Taiwan, especially in the fields of natural science and electronics-related engineering. Every year, hundreds of highly-talented young engineers graduate from these two universities and join high-tech industries in the Park. The Park was located within the vicinity of these academic and research institutions in order to ensure that the Park-based companies would have easy access to a talented pool of young engineers and to other technological resources needed. Therefore, the synergy created by the concentration of academics, researchers and industry has contributed a great deal to the outstanding growth of Park-based companies and high-tech industry in Taiwan more generally.

Third, returning expatriates have played a vital role in the rapid growth of PBEs. By the end of 1995, 2,080 persons had returned from abroad to work in the Park. They came back either to open their own businesses or to join newly founded companies. The number of companies founded by these returnees was 79 in 1995, which accounted for more than 40 per cent of all Park-based enterprises. Their valuable contributions included new ideas, technology, entrepreneurship and experience they brought back to Taiwan. For example, in 1980, an expatriate came

Table 4.11 HSIP: status of high-tech industry, 1995

Industry	*Firms*	*Employees*	*Sales (US$ million)*	*Growth (%)*
Integrated circuits (ICs)	56	22,496	5,591	76
Computers and peripherals	41	11,148	4,593	69
Telecommunications	30	4,071	643	15
Opto-electronics	26	3,270	379	112
Precision machinery and materials	18	1,041	94	28
Bio-technology	9	231	8	–46
Total	180	42,257	11,308	68

Source: HSIP, *Annual Report 1995*.

back from the United States to start the first scanner manufacturing firm in Taiwan. Over the years, this company has grown to become one of the leading manufacturers in the industry, while many other scanner producers have been directly or indirectly helped by the pioneering company. Taiwan is now the number one scanner producer in the world, with a market share in 1996 of 61 per cent. Such stories are quite common in the Park.

As shown in Table 4.11, almost one-third of the PBEs are in the IC industry, with 41 in the computers and peripherals industry. The combined sales of these industries accounted for 90 per cent of total sales from the Park, with half of them from the IC industry. Although all officially designated high-tech industries are encouraged in the Park, only the information technology industry has flourished.

Taiwan's IT industry conditions

The fact that only the information technology industry has become very competitive in the world market reminds us that in order to develop an industry, an economy must possess some critical economic factors. We use a paradigm developed by Michael Porter to explain why the information industry in Taiwan became so internationally competitive in such a short period of time. According to Porter (1990), in order to achieve international success in a certain industry, an economy has to possess competitive advantages in the following four regards: (1) factor conditions such as capital, technology, manpower and infrastructure; (2) demand conditions, domestic and abroad; (3) related and supporting industries; and (4) firm strategy, market structure and rivalry. These conditions are reviewed below. We will also evaluate the role played by the government in setting up certain institutions and economic infrastructure for enhancing the competitiveness of the industry.

Factor conditions

In order to develop an internationally competitive industry, a country must have a strong position in at least one of the following four production factors: capital, skilled labour, technology and management. When the economy began to emphasise more technology and capital-intensive industries in the early 1980s, Taiwan's economic environment was transformed for the challenge. First, capital was not a problem for the development of the information industry. The economy had been growing very rapidly since the 1960s, with the savings rate in Taiwan high but still rising. Through formal or informal channels, these savings supplied ample investment funds for capital-intensive industries. Since venture capital was not very popular at that time, the government was very active in putting up investment capital for designated high-tech industries. Sometimes, besides making available low-interest loans, the government invested directly or got government-owned banks, such as the Chiao-Tong Bank, to invest.

However, most small and medium enterprises have had to rely on the personal resources of the entrepreneur for their capital. Hence, personal wealth was a critical source for a new start-up company. For example, Acer, a leading firm in Taiwan's computer industry with annual sales of more than US$3 billion in 1996, was founded with US$250,000, with all the capital coming from the personal savings of the co-founders, relatives and friends. The rapid growth of these high-tech companies made it necessary for more capital to be put into the industry. When personal savings proved inadequate, these companies had to turn to the capital market to get the money needed for expansion. In fact, most important producers in Taiwan's information industry are listed in the stock market. Of the total of 180 firms in the Park, more than 10 per cent are already listed in the stock market. The high growth potential of these high-tech firms made it very common for their prices to quickly rocket to more than double or triple the originally quoted prices.[11] This has made it easier for other high-tech firms to find needed capital in the market.

Second, by the 1980s, Taiwanese enterprises had accumulated considerable experience in producing electronic commodities. During the 1960s and 1970s, many American and Japanese multinationals came to Taiwan to set up offshore production bases for consumer electronics such as TVs, calculators, video games, and a variety of parts and components. All these products have consisted of many components and parts, and have been produced on a large scale to be competitive. Hence, competitiveness derives from the speed and costs of production. Usually, a firm has to organise production very efficiently and flexibly to be competitive. To do so, a firm not only has to have very good management skills, but also has to attract and keep skilful and hard-working workers to work on the assembly line.

Taiwan already had a lot of foreign direct investments in the IC packaging and other electronic component industries in the 1960s. General Instruments, Philips, Texas Instruments and RCA were among the many multinationals that came to Taiwan to set up their semiconductor assembly operations. By the late 1960s, some domestic firms had started to emerge with their own IC assembly businesses. But these domestic firms were relatively small in scale and relied heavily on subcontracting work from foreign firm subsidiaries in Taiwan. In 1971, the first domestic IC manufacturing firm was formed with assistance from the semiconductor centre based in Chiao-Tong University. In the end the firm failed, but the experience gained was valuable, with those who had worked in the firm becoming key persons in Taiwan's subsequent successful development of the IC industry.

Third, the contribution of direct foreign investment (DFI) to Taiwan's economy is well documented in the literature. Foreign investments in the manufacturing sector were 32, 187 and 713 million US dollars in the 1960s, 1970s and 1980s respectively, with a great proportion of these investments in the electronics industry. These investments contributed greatly not only to capital formation, employment and exports, but also to technology transfer.[12] Such technology transfer was largely embodied in people who served as supervisors, technicians and managers in foreign subsidiaries. In the 1960s and 1970s, foreign subsidiaries were the first choice for young university graduates because they usually provided better working conditions and higher pay. These people acquire experience and knowledge of production while working for foreign firms. Many important people in Taiwan's PC industry have had close relations with these foreign subsidiaries, while their experience and knowledge have greatly contributed to the development of Taiwan's PC industry.

In the past two decades, many workers have returned to Taiwan to start their own businesses or to join the booming high-tech industries. These returned expatriates have contributed to Taiwan's information technology industry by bringing back their knowledge of advanced technology and experience. Expatriates have founded some eighty companies in the Hsin-chu Science-based Industrial Park, which account for 40 per cent of all PBEs, while the number of expatriates working in the Park exceeds 2,000. These people have been very important to the development of Taiwan's high-tech industry because most have had many years of experience working for leading firms in the United States.

Fourth, high-quality and well-disciplined workers have been another important factor in Taiwan's economic miracle. For the past twenty years, the educational system has trained sufficient university graduates to fulfil the needs of the booming information technology industry. According to government statistics, the total number of four-year university graduates – including first (Bachelor) and higher degrees – was 80,828 in 1995. Around one-third of these graduates hold science and engineering degrees; 3,972 graduates were from information technology-related departments. If we take colleges into account, the number of graduates increased to 181,621, and the ratio of those with engineering and science backgrounds rose to more than 40 per cent. These statistics suggest that the supply of talented and skilled labour has not been a constraint to the development of the information technology industry in Taiwan.

Fifth, as noted earlier, the government has been actively involved in the development of technology for high-tech industries. ITRI was established at the very beginning of the development of Taiwan's electronics and information industry. Its main goal is to serve the technological needs of Taiwan's industrial development. Besides its role in the development of IC technology, ITRI has extended its activities to cover chemicals, machinery and the aerospace industry, all major industries in Taiwan. The Institute of Information Industry (III) has been important for the promotion of computer programming, systems technology and computer usage. Besides institutional help, the government also provides grants for the private sector to develop new technology and new products. Tax incentives are also available for domestic enterprises, with R&D expenses deductible for income tax purposes.

Demand conditions

The PC industry is so dynamic that new products are coming on the market all the time. Since the life cycles of many PC-related products have become extremely short, a firm has to be very productive in the sense that the product is manufactured quickly and cheaply in order to recoup expenses in developing a new product before competitors can catch up. In such a dynamic environment, critical events in the world PC market have helped Taiwan's PC industry to emerge.

In the early 1980s, IBM created an 'open architecture' which allowed PC makers throughout the world to produce IBM-compatible PCs. This decision greatly lowered entry barriers for Taiwan's PC producers, both financially and technologically (Kraemer *et al.*, 1996). As most Taiwanese producers were very small, they could not afford to launch their own brand products in the market. Therefore, the opportunity of producing PCs for IBM and other big name PC firms was just what Taiwan needed at that time. Most Taiwanese PC firms still rely on work as original equipment manufacturers (OEM) or original design manufacturers (ODM) for big foreign PC firms.

In early 1992, Compaq, one of the leading computer vendors in the world, led the world PC market into a fierce price war. The price war meant that production costs mattered to all leading PC companies. To remain competitive in the market, the big players in the PC market had to out-source their production elsewhere to cut costs. Once again, Taiwan was a 'natural choice' for these firms.

Compared with the giant Korean and Japanese electronic producers, Taiwanese firms are much smaller in size and less diversified. However, with a very competitive industrial structure, composed mostly of very agile small and medium-sized enterprises, Taiwan has been able to surpass them in the production of several PC-related products. The much smaller Taiwanese firms have been able to beat their bigger counterparts in Korea and Japan because they have been more adaptive to the rapidly changing PC markets. Simply put, smaller firms in Taiwan can more easily adjust to changes in the market and have therefore been better able to find new niches in the PC industry's international division of labour, compared to bigger companies.

Related and supporting industries

As shown in Table 4.5, Taiwan has the most complete line of competitive computer-related industries in the world. Taiwan's enterprises have a very strong competitive advantage for all computer peripherals and components, except for the production of CPUs and hard-disk drives. By the late 1990s, Taiwan was the world's number one maker of monitors, keyboards, motherboards, power suppliers, mice, scanners, graph cards, and LAN cards. Their competitiveness greatly supports PC makers in Taiwan. In fact, some producers of peripherals are affiliated to PC makers. For example, Acer is itself a computer maker, but the Acer group also designs and makes monitors, motherboards and even ICs for others.

Recently, off-shore production of computer peripherals has become much more important than a decade ago because production costs have increased so much that domestic firms have had to move some of their production processes or even entire product lines abroad to cut costs. Almost all the leading firms in Taiwan's PC industry have set up overseas production lines to cope with the problem of rising production costs in Taiwan. By 1996, off-shore production accounted for more than one-third of Taiwan's total production of PC-related information technology products. As shown in Table 4.5, almost all keyboard production has already gone overseas as well as two-thirds of mouse output, half the monitor production, and over one-third of motherboard manufacturing. With this outward movement, domestic firms have moved into the production of higher-end products and/or to expand their production scale to further reduce their costs.

In the case of the IC industry, Taiwan did not have its own IC manufacturing firm until UMC and TSMC went into operation with strong support from the government. Many domestic enterprises followed their success and went into designing, packaging, testing and photo-masking ICs for them. Their strategy has been to specialise in particular areas of the IC industry. Just as the TSMC specialises in foundry services, some other domestic firms specialise in packaging, photo-making, etc., thus growing together with the two big IC manufacturers in Taiwan.

Firm strategy

As is well known, industrial structure in Taiwan is characterised by its very dynamic and competitive small and medium-sized enterprises. PC-related industries are no exception. SMEs in these industries are very competitive because the export-promotion policies implemented since the early 1960s have exposed them to the world market, where only the most competitive firms survive. Their ability to compensate for their size disadvantage by forming closely-knit production networks among themselves has also contributed to their competitiveness. Being SMEs, most Taiwanese companies can only attain competitive advantage by specialising in producing only one or a few product items. Connected to well-functioning networks, these SMEs have become formidable competitors in the world market.[13]

Nowadays, there are only a few leading brand names in the world PC market. As can be seen in Table 4.7, of the top ten PC suppliers, which accounted for more than 50 per cent of the world market, six were American companies, three were

Japanese, with Taiwan's Acer the only one from a developing country. Since most Taiwanese firms are too small to have their own brand name products in the market, they adopt the so-called 'fast follower' strategy. The adoption of such a strategy allows them to concentrate on a portion of the value chain. Thus, Taiwanese firms continuously upgrade their manufacturing efficiency where they have competitive advantage, and leave pioneering R&D and marketing to the leading PC firms. By doing so, Taiwan's enterprises have gained strong positions in international commodity chains. However, this does not mean that Taiwanese firms do not innovate. As shown in Table 4.10, the firms in the Park spent 4.6 per cent on R&D activity on average in 1994. Besides, to keep abreast of new technology advances, most Park-based enterprises either have opened a branch office or already own an affiliated company specialising in R&D in the United States (most of them are located in Silicon Valley).

By focusing on components, with very few gradually expanding into the markets for final products such as PCs, Taiwanese firms have enjoyed some advantages. First, the production of components is relatively easier compared with the production of final products while the capital needed is smaller as well. Second, to sell a final product in consumer markets requires considerable marketing capabilities in alien markets which Taiwanese firms do not expect to be good at. Third, unlike final products, components are usually designed by the final product maker so that a component maker does not have to bear too much R&D expense. For these three reasons, Taiwanese firms almost always choose to start their businesses as OEM producers for foreign firms, and then only gradually progress up to producing more sophisticated products.

Conclusion

In this chapter, we have tried to explain why Taiwan's information technology industry has developed into such an internationally competitive industry in such a short period of time. In Porter's paradigm, a nation has to possess four conditions to gain strong international competitiveness in an industry. Although the four elements are all subject to market forces, government policy does have a role to play to enhance and/or supplement market forces.

Our brief review of Taiwan's successful economic development and various policies implemented in the past four decades emphasised how the economy has been transformed from a relatively labour-intensive industrial structure into a more technologically and capital-intensive structure. In every stage of this transformation, the helping hand of government has guided domestic enterprises to invest in industries in which Taiwan quickly developed comparative advantage. From the early 1960s, the government has adamantly insisted that the market should be as competitive as possible so that the private sector has to contend with market forces and fight for survival through fierce competition. As the entry barrier is quite low for most industries, a lot of small and medium-sized firms enter markets when they see niches and exit when profit margins are too thin to sustain operations. Thus, only those with the greatest competitive advantage can survive and grow.

The role of government has evolved with changes in Taiwan's comparative advantage. It has become more sophisticated and more involved in the development of industry as the economy has become more industrialised. For example, when labour costs in Taiwan were relatively low in the 1960s, the government implemented policies to encourage local enterprises to invest in labour-intensive industries. But, as the economy turned to more technology- and/or capital-intensive industry, the government has helped develop key technologies for the private sector, e.g. industrial policy for the development of the information technology industry since the 1980s.

As discussed above, for the development of the information technology industry, the government has adopted more than just a market-oriented open economic strategy. ITRI and III were created to help develop key technology for the industry, to provide technical services for the private sector to surmount bottlenecks. Sometimes, as in the case of the IC industry, the government has had to put up capital for spin-off firms such as the UMC and the TSMC because the private sector saw the risks as too high and would not invest. These two institutions also trained many engineers and technicians through government-funded research projects. These well-trained engineers and technicians are usually highly demanded by the private sector. The establishment of the HSIP was another institutional initiative for the birth and growth of designated high-tech industries in Taiwan. It has been so successful that a second one has been constructed in the southern part of Taiwan to accommodate the rapid growth of the information technology industry.

The contribution of these three institutions to the successful development of the information technology industry in Taiwan has been crucial. These institutions have not been designed only for the information technology industry; other high-tech industries have also been promoted by ITRI and the HSIP. But only the information technology industry became internationally competitive in the world market. Some elements distinguish the information technology industry from other industries. Taiwan's industrial structure and very active small and medium-sized enterprises have put it in a strong position in the international personal computer industry. Although Taiwan does not have many big names in the industry, Taiwanese SMEs built up competitiveness by specialising in particular computer-related products. Their strongest point has been their flexibility and dynamics crucial for today's very competitive world computer market. As indicated earlier, for several computer-related products Taiwan's market share is well over 50 per cent. Taiwan's many very competitive producers have made it a very important international manufacturing centre of computer products.

Taiwan has been able to take good advantage of foreign technology, either through the multinationals operating in Taiwan since the early 1960s, or through expatriates who have studied and worked in the United States. Both multinationals and expatriates have contributed greatly to transferring advanced technology, new management skills and fresh ideas to Taiwan. Many expatriates returned to Taiwan to either work in or run high-tech enterprises in the Hsin-chu Science-based Industrial Park. Even some who stay in the United States work for Taiwanese firms

– e.g. in product development or marketing – and have been crucial for the development of the information technology industry.

Various challenges lie ahead. First, as the economy becomes more liberalised and internationalised, the government's ability to design and implement effective and timely industrial policy has been eroding. However, as the private sector becomes much more competitive, stronger and bigger, the need for guidance, support and assistance from the government is also eroding. There has been some strong criticism of ITRI and III for their roles in taking too many government R&D resources, and for alleged inefficiency and inability to fulfil the needs of the private sector. Nevertheless, demand for the Hsin-chu Science-based Industrial Park remains high because the government provides better and more efficient services in the Park than outside.

Second, the 'fast follower' strategy may backfire someday because Taiwan is too technologically dependent on both American and Japanese companies for key components such as CPUs, LCD panels, hard-disk drives, high-resolution tubes, and IC production equipment. As followers, Taiwanese firms do not have to spend much on R&D and marketing. Hence, their competitive advantage largely depends on production efficiency, which can be rather easily replicated by other latecomers. Therefore, there is a risk that leading PC firms may decide to out-source their production elsewhere when Taiwanese firms are no longer able to produce cost-competitively for them.

Notes

1 Only in rare cases (such as the two oil crises) did price levels increase by double digits.
2 See Wu (1991) for a discussion of how the government has helped the economy to adjust to the changing international environment.
3 According to the statistics of host countries, these numbers are greatly under-estimated. The sum of reported investment from Taiwan by Thailand, Malaysia, Indonesia, Philippines and Vietnam amounted to US$32 billion up to 1996, while investments in China were more than US$30 billion in the ten years before 1997.
4 By the IT industry, we are mainly referring to the hardware part of the industry, such as personal computers (PCs), computer peripherals, components and integrated circuits (ICs).
5 Taking overseas production by Taiwanese firms into account, the amount is around US$30 billion.
6 Toshiba is still the world's largest portable computer producer.
7 In 1995, Acer had 15,000 employees, 80 offices in 38 countries, and dealers in 100 countries.
8 The HSIP was modelled on world-famous specialised industrial parks such as the Research Triangle in North Carolina and Japan's Tsukuba Science City.
9 The National Science Council is the highest ranking authority for Taiwan's science policy.
10 If we take R&D financed by the government out of the total amount of R&D expenditure, then the ratio becomes much less than 2 per cent.
11 In terms of market value, both UMC and TSMC are now among the ten largest firms in the stock market.
12 DFI accounted for 11.06 per cent, 10.94 per cent and 5.84 per cent of domestic capital

formation in the 1960s, 1970s and 1980s respectively. See Tsiang and Wu (1985); Wu (1989); Ranis and Schive (1985).

13 For example, subcontractors receiving large orders can reduce production costs significantly due to scale economies. For example, the cost of producing a power supply can be reduced by 2 to 3 dollars each if monthly production can be increased to 50 thousand units.

References

Aoki, Masahiko, Kevin Murdock and Masahiro Okuno-Fujiwara (1996) 'Beyond the East Asian miracle: introducing the market enhancing view', Washington, DC: Economic Development Institute, World Bank.

Deyo, Frederick C. (1987) *The Political Economy of the New Asian Industrialism*, Ithaca, NY: Cornell University Press.

Galenson, Walter (ed.) (1979) *Economic Growth and Structural Change in Taiwan*, Ithaca, NY: Cornell University Press.

—— (ed.) (1985) *Foreign Trade and Investment: Development in the Newly Industrialized Asian Countries*, Madison, WI: University of Wisconsin Press.

Gereffi, Gary and Donald L. Wyman (eds) (1990) *Manufacturing Miracles: Paths of Industrialization in Latin America and East Asia*, Princeton, NJ: Princeton University Press.

Kim J.I. and L.J. Lau (1994) 'The sources of economic growth in the East Asian newly industrialized countries', *Journal of the Japanese and International Economies*, 8: 235–71.

Kraemer, K.L., J. Dedrick, Hwang Chin-Yeong, Tu Tze-Chen and Yap Chee-Sing (1996) 'Entrepreneurship, flexibility and policy coordination: Taiwan's computer industry', *The Information Society*, 12: 215–49.

Lau, Laurence J. (ed.) (1990) *Models of Development: A Comparative Study of Economic Growth in South Korea and Taiwan*, San Francisco: Institute for Contemporary Studies.

Lee, K.T. (1988) *The Evolution of Policy Behind Taiwan's Development Success*, New Haven, CT: Yale University Press.

Levy, B. and W.J. Kuo (1991) 'The strategic orientation of firms and the performance of Korea and Taiwan in frontier industries: lessons from comparative case studies of keyboard and personal computer assembly', *World Development*, 19(4): 363–74.

Porter, Michael E. (1990) *The Competitive Advantage of Nations*, London: Macmillan.

Ranis, Gustav (1979) 'Industrial development', in Walter Galenson (ed.) *Economic Growth and Structural Change in Taiwan*, Ithaca, NY: Cornell University Press.

Ranis, Gustav and Chi Schive (1985) 'Direct foreign investment in Taiwan's development', in Walter Galenson (ed.) *Foreign Trade and Investment: Development in the Newly Industrialized Asian Countries*, Madison, WI: University of Wisconsin Press.

Rushing, Francis W. and Carole G. Brown (eds) (1986) *National Policies for Developing High Technology Industries: International Comparisons*, Boulder, CO: Westview.

Schive, Chi (1990) 'The next industrialization in Taiwan and South Korea', in Gary Gereffi and Donald L. Wyman (eds) *Manufacturing Miracles: Paths of Industrialization in Latin America and East Asia*, Princeton, NJ: Princeton University Press.

Simon, Dennis F. and Chi Schive (1986) 'Taiwan's information industry: the role of the state in the development of high-tech industries', in Francis W. Rushing and Carole G. Brown (eds) *National Policies for Developing High Technology Industries: International Comparisons*, Boulder, CO: Westview.

Tsiang, S.C. (1984) 'Taiwan's economic miracle: lessons in economic development', in Arnold C. Harberger (ed.) *World Economic Growth*, San Francisco, CA: ICS Press.

Tsiang, S.C. and Wu Rong-I (1985) 'Foreign trade and investment as boosters for take-off: the experiences of the four Asian newly industrialized countries', in Walter Galenson (ed.) *Foreign Trade and Investment: Development in the Newly Industrialized Asian Countries*, Madison, WI: University of Wisconsin Press.

Vogel, Ezra F. (1991) *The Four Little Dragons: The Spread of Industrialization in East Asia*, Cambridge, MA: Harvard University Press.

Wade, Robert (1990) 'Industrial policy in East Asia: does it lead or follow the market?', in Gary Gereffi and Donald L. Wyman (eds) *Manufacturing Miracles: Paths of Industrialization in Latin America and East Asia*, Princeton, NJ: Princeton University Press.

Wang Zuo-Rong, Wu Rong-I, *et al.* (1988) *Review on Economic Development Policy and Systems in Taiwan* (in Chinese), Taipei: Council for Economic Planning and Development, Executive Yuan.

World Bank (1993) *The East Asian Miracle: Economic Growth and Public Policy*, New York: Oxford University Press.

Wu Rong-I (1989) 'Foreign direct investment and technology transfer', paper presented at a conference on 'State Policy and Economic Development in Taiwan', Taipei, Taiwan, 4–5 December.

—— (1991) 'Taiwan: adjustment in an export-oriented economy', in Hugh Patrick (ed.) *Pacific Basin Industries in Distress: Structural Adjustment and Trade Policy in the Nine Industrialized Economies*, New York: Columbia University Press.

Wu Rong-I, *et al.* (1986) *An Evaluation of the Performance of the Industrial Technology Research Institute* (in Chinese), Tai-chung: National Chung-Hsing University.

5 Industrial policy and the emergence of internationally competitive manufacturing firms in Malaysia

Jomo K.S., Rajah Rasiah, Rokiah Alavi and Jaya Gopal

In recent years, the factors underlying East and Southeast Asian economic growth have been increasingly discussed in the development literature, with economists offering various explanations for this success. Amsden (1989) argues that South Korea's spectacular industrial achievement can be explained by government subsidies, tariff and non-tariff incentives, financial credit facilities, a highly educated and trained workforce, firm capabilities to learn and adapt foreign technology, and the government role in linking incentives to time-bound (limited period) performance standards (especially exports). Even the World Bank (1993) now recognises the role of government intervention in spurring industries to export. Unlike in Southeast Asia, industrial growth in East Asian economies was largely led by indigenous firms. In Southeast Asia, however, foreign firms have dominated much of export manufacturing.

Amsden suggests that firms in late industrialising countries go through four stages in becoming successful learners. First, they compete to get industrial licences and contracts from the government. Second, they compete to get foreign technical licences from international firms on the best available terms. Third, they compete in the labour market for the best recruits, supervisors, managers and engineers, in terms of experience and skills. Fourth, they compete in the market place on the basis of cost, quality and reliability. Amsden and Kim (1985) have suggested that the forms of technology acquisition have changed over time, from the earlier tendency for firms to absorb foreign technology through copying and learning on their own, to adapting foreign technology after investing in foreign licences and technical assistance. The former mode of technology acquisition may be called imitation, and the latter, apprenticeship, i.e. learning by doing. Another important mode of technology acquisition is by buying the firm that owns the technology.

Jones and Sakong (1980: 81) see a successful entrepreneur as one able to become competitive as a consequence of government intervention. The tasks of an entrepreneur, according to them, include the following:

1 perception of a new economic opportunity, including a new product, a new production process or a new market
2 evaluation of the profitability of a new opportunity
3 gaining command of financial resources
4 supervision of plant design, technology and construction
5 recruitment and training of new personnel
6 good relations with government
7 good relations with suppliers and purchasers.

As has been demonstrated by Schumpeter (1934), Khan (1989) and Chang (1994), rents are necessary to stimulate innovative investments that are risky and lumpy.[1] Powerful interest groups can emerge independently of the state to distort the allocative capacity of markets (Kornai, 1979; Rasiah, 1995: chapter 2). For such reasons, Khan (1989), Chang (1994) and Rasiah (1996b) have argued that governments creating such rents should ensure that they are temporary and utilised productively. State governance, however, does not necessarily guarantee effective appropriation of rents. Even non-corrupt bureaucrats who possess little knowledge of markets and technology are prone to failure.

To minimise government failures, effective co-ordination through markets becomes essential. As markets are generally underdeveloped in developing economies, contrary to common assumptions, e.g. by Krueger (1974), Bhagwati (1988), Lucas (1988) and Helpman and Krugman (1989), they cannot be allowed to dictate allocation. Both state and market are susceptible to influence by powerful interest groups. Since unproductive collusion between powerful interests from both sides will fetter accumulation, pro-active, dynamic, growth-oriented co-operation between the two have been important for successful rapid growth as the experiences of Japan, Korea and Taiwan have shown. Given their respective strengths and weaknesses, state and market should complement each other to efficiently co-ordinate the creation and allocation of rents (Rasiah, 1995, 1996b).

Schumpeter has argued that various restrictive practices may increase profits and reduce the risks faced by firms that undertake the costly investments required for innovation. Schumpeter (1975: 102) explained that 'a monopoly position is in general no cushion to sleep on. As it be gained, so it can be retained only by alertness and energy'. Ekelund and Tollison (1981: 18–19) acknowledge that rents provide the incentive for resource owners to seek out more profitable (and, presumably, more economically efficient) allocation of their resources. While there is a tendency for rent-seeking to result in unproductive, corrupt and wasteful activities in politically modified markets, state intervention can also reshape growth and accumulation processes to facilitate the emergence and development of new economic activities (see Schumpeter, 1975: 78; Chang, 1994; Jomo, 1996: 5; Khan, 1996).

Hence, the prospect of capturing rents stimulates entrepreneurial decisions, e.g. to invest in research and development to bring about technological change in the Schumpeterian sense, and can also be presumed to bring about a correspondingly efficient allocation of resources. Even rent-seeking welfare losses may well be more than offset by the dynamic gains of productivity growth which the rent facilitates, e.g. by increasing opportunities for learning by doing, as in the case of infant industry protection, or by encouraging firms to spend more on research and development. Hence, while rent-seeking may be directly unproductive, it may well constitute transaction costs which indirectly facilitate productivity gains (Chang, 1994). Such distortions are recognised, for example, to have been important in facilitating late industrialisation in continental Europe, the US and Japan in the nineteenth century.

With increasing globalisation and trade liberalisation, achieving international competitiveness has gained considerable significance and attention in both developing and developed countries (Porter, 1990; UNIDO, 1995). Citing empirical evidence, Katz (1984) Fransman (1986), Nelson (1987), Lall (1992), and others have criticised the static neo-classical framework and argued that a more dynamic approach is necessary for analysing the comparative advantage of industrial projects and industries in developing countries. Others have also argued for the need to consider technical change and technological capacity in industrial project evaluation (Fransman, 1982: 1008–9; Bell *et al.*, 1984: 102–3; Weiss, 1986: 173–4; 1989: 496–505) and of the positive role of intervention in achieving technological learning and competence (Lall, 1992, 1995, 1996). However, outside the East Asian NICs, there has been limited evidence of firms/industries achieving international competitiveness (Bell *et al.*, 1984: 111–14, 123; Weiss, 1986: 172; Herbert-Copley, 1990: 1463).

Doner (1992), however, argued that entrepreneurship has also been important in making local companies internationally competitive. In stressing the importance of institutional factors in the development and success of many local business groups, he argues that many writers have neglected the role of non-governmental institutional factors – such as business groups, business-interest associations, networking systems and the relationship between the government and the private sector. Suehiro (1996) adds that political connections alone cannot always determine or guarantee the success of a firm and, therefore, the rapid expansion and growth of specific business groups cannot be simply attributed to connections with the government or collaboration with foreign firms alone. Other factors have helped many domestic private firms to advance and develop, e.g. managerial skills, technological innovation, marketing and other capabilities.

The factors behind a firm's success are likely to be complex and manifold. In trying to understand the conditions in which internationally competitive Malaysian manufacturing firms have emerged in Malaysia, we begin by reviewing the economic, especially the policy environment, in which firms have achieved success, whether real or apparent. Our introductory review of Malaysian industrialisation suggests that much of the growth of the manufacturing sector has been due to policy interventions of the post-colonial government. While export-oriented firms seem to be much more internationally competitive, they too have benefited considerably from selective industrial policy.

This is followed by four case studies – three of firms and the other of an industry.

Eng Hardware's experience highlights the potential for indigenous supplier firms to emerge in the foreign dominated electronic components industry. Sapura's case is probably the most successful example of a politically favoured private *Bumiputera* manufacturing firm that has developed on the basis of rents allocated by the state on an ethnic basis. Our third case study of Proton, the (first) national car industrial firm, offers a story of mixed successes. Finally, the successful emergence of the palm oil refining industry in Malaysia is probably the greatest achievement of selective industrial policy in Malaysia. Taken together, these cases offer a more nuanced view of the role of industrial policy in Malaysian industrialisation than has been the case so far.

State intervention, rents and Malaysian industrialisation

Being a fast growing economy, Malaysia is often classified as liberal, with its export-oriented industries considered to be governed by free markets (Sheperd, 1980; World Bank, 1993). The international competitiveness of the export-oriented sub-sector appears to be central to arguments about Malaysia's success in achieving fast growth. Export-orientation is often equated with market-orientation. Indeed, Sheperd (1980: 186–7) has argued that the rapid growth of export-oriented industries was achieved through competitiveness, not subsidies.

Evidence of the interventionism of the governments of South Korea, Taiwan and Singapore is now clear (Luedde-Neurath, 1986; Evans, 1987; Amsden, 1989; Wade, 1990). It will be argued here that Malaysia's import-substituting (IS) and export-oriented (EO) manufacturing sectors have both been regulated, with relative prices distorted. The IS and EO sectors have both enjoyed active state promotion, with IS sectors protected and EC sectors subsidised. Growth of the import-substituting sector became sluggish as protection has not been offset by other interventions to ensure technological dynamism. Distortions created by the government have enabled transnationals producing for global markets to benefit from various host economy incentives for the EO sector.

Foreign capital contributes a significant proportion of fixed assets, employment and output in the manufacturing sector. Except for leather, wood and basic metals, foreign capital owned more than half of fixed assets in the remaining manufacturing branches in 1968. This gives a rough indication of foreign ownership in Malaysia's manufacturing sector until the 1970s. The modern manufacturing sector which emerged during colonialism began to expand after independence in 1957, with foreign firms setting up assembly, packaging and other finishing manufacturing operations to benefit from the high tariffs introduced following the Pioneer Industries Ordinance of 1958. This demand–pull effect on foreign capital began to fall gradually as the local market became saturated.

The next wave of foreign relocation came after the Free Trade Zone Act of 1971. This Act attracted mainly export-oriented labour-intensive industries. The electric/electronic, textile/garment and rubber industries were the main beneficiaries of this thrust. Rapidly growing domestic demand in the early 1970s, especially in the agricultural and transport sectors, led to the expansion of foreign participation in chemicals, machinery and transport equipment. During 1975–9, foreign ownership declined relatively in almost every manufacturing branch, while foreign manufacturing fixed assets grew less in this period. Wood manufacturing was the only branch with a clear increase of foreign ownership, but this branch was relatively insignificant. The initial relocation of labour-intensive firms from developed economies slowed down in this period,[2] while the state offered few improvements to existing incentives.

The overall proportion of foreign ownership in manufacturing fell gradually in the 1970s, except in 1971–4, when it was stable. The foreign share in manufacturing fell further until the mid-1980s despite a rapid increase in foreign investment in the 1980s, mainly due to a dramatic rise in local investment, especially in state-sponsored heavy industry. Nevertheless, in the second half of the 1980s, the relative share of foreign capital, especially from Japan and the Asian NICs, rose again following increasing trade restrictions imposed on their exports by governments from the major markets of the West. Privileges under the Generalised System of Preferences (GSP) were withdrawn from the Asian NICs in 1988. Malaysia was an important beneficiary of such investments, which explained the rising trend in foreign ownership of fixed capital in most manufacturing industries.

As with most newly independent economies, Malaysia adopted an import-substitution strategy following the Pioneer Industries Ordinance of 1958 to promote industrialisation. Unlike the experience of more nationalistic regimes, such as South Korea and Taiwan, local capital was not given a leading role. Industrial policy during the import-substitution phase neither prioritised local firms nor discriminated against foreign capital. Both foreign and local firms enjoyed similar incentives in the industries promoted. With import substitution being the main industrialisation strategy in the 1958–68 period, foreign firms relocated many industrial operations to benefit from high tariffs. Indeed, foreign firms dominated ownership of most import-substituting industries in the 1960s. Both the modern textile and electronic industries first emerged during this phase; a textile factory was first established in 1957, while a joint-venture electronics plant was started in 1967. Unions were also discouraged in new industries following the Pioneer Industries Ordinance of 1958 and the Trade Unions Act of 1959.

IS had been prescribed by structural economists, such as Lewis (1955) and Myrdal (1957), to develop infant industries for eventual international competitiveness. However, while the Malaysian government used import-substitution incentives to attract foreign firms, it did not seem to see import substitution as a means for eventual export promotion.[3] Foreign firms from Singapore and Britain in particular relocated 'screw-driver' assembly operations to benefit from the high tariffs imposed on finished goods (see Edwards, 1975; Saham, 1980). As raw materials and intermediate goods generally faced low tariffs, import-substituting firms were highly protected in the 1958–68 period. Most foreign firms expanding operations in Malaysia during this phase also had production operations elsewhere, which generally discouraged exports from Malaysia. Besides, pioneer status incentives were only offered for import substitution during this decade. Thus, when the small domestic market became saturated by the mid-1960s, there appeared to be little additional demand to stimulate expansion. Manufacturing's contribution to gross domestic product (GDP) in 1960 and 1965 therefore stagnated at 9 per cent (World Bank, 1980).

Although the high protection reduced competition in the import-substitution phase, protection without efficiency-inducing incentives largely accounted for the eventual stagnation of manufacturing during the import-substitution phase. While the state in South Korea offered import-substitution rents to local capital in return for meeting stringent performance standards (Amsden, 1989; Chang, 1994), the Malaysian state offered such rents to mainly foreign firms without imposing performance conditions, i.e. the carrot was given without the stick. In South Korea, the state simulated competition to press for technical change and efficiency improvements as new firms could hardly be expected to compete with foreign transnationals from the outset. Hence, the South Korean government succeeded in creating dynamic comparative advantage by encouraging import substitution only as an initial step toward export-orientation. The Malaysian state did not similarly use import substitution progressively, which could have been due to a weak sense of economic nationalism due to the state's ethno-populist priorities.

Following the shift to export-orientation after 1968, import substitution gradually lost significance in terms of output and employment generation, although it continued to coexist with the former. Import substitution also declined in significance as several tariffs gradually fell, thereby reducing the distortionary rents enjoyed by these industries. For example, between 1969 and 1987, the effective rate of protection (ERP) for basic industrial chemicals fell from 160 per cent to 16 per cent, for tobacco from 125 per cent to –26 per cent, for fertilisers and insecticides from 300 per cent to 8 per cent, and for structural metal products from 35 to 1 per cent.[4] While EO firms continued to enjoy generous incentives, tariff protection for most import-substituting firms gradually declined.

However, some import-substituting industries, earmarked for strategic promotion by the government, experienced increases in their ERP; e.g. the ERP for basic iron and steel rose from 28 per cent in 1969 to 131 per cent in 1987. From 1981, the Malaysian government intervened strongly to promote heavy industries in Malaysia, through the Heavy Industries Corporation of Malaysia (HICOM). This was not merely second-order import substitution as the key feature of this strategy appears to have been direct government participation in developing heavy industry. Earlier import substitution had concentrated on attracting import-substituting investment irrespective of ownership, control, structural content or scale, while the early 1980s saw the introduction of state-sponsored and controlled investment in heavy industry. These heavy industries enjoyed top priority from the government and were not integrated with the earlier import-substitution sector. Moreover, the second import-substitution phase involved few structural links with the EO sector. Hence, there was no programmatic sequencing of industrial policy typical of other second rounds of import substitution.

In addition to protection, the government also offered subsidised capital, imposed stringent controls on competitors in the domestic market and introduced other promotional tools to spawn the manufacture of cement (Kedah Cement and Perak Hanjoong), steel (Perwaja Steel) and motorcars (Proton). The government's objectives included, *inter alia*, the development of a strong capital goods sector and linkages with the domestic economy, especially *Bumiputera* enterprises. These industries remain strongly subsidised and protected. Indeed, by controlling Proton's purchases, the government has been gradually requiring domestication of parts supply and *Bumiputera* participation involving an 'umbrella' approach to vendor development. In this way, by 1993, Proton had achieved domestication of 80 per cent of car components, and forced firms supplying Proton to raise *Bumiputera* equity participation. Despite its import-substitution origins, the automobile industry was quickly reoriented to exports, albeit the share of exports has remained very small. The monopoly rents enjoyed in the domestic market due to high tariffs effectively subsidise exports. While elements of IS-for-EO exist for these heavy industries, it is unclear if such rents will be gradually withdrawn with rising sales volumes and competitiveness.

Except for beverages and tobacco, domestically-oriented industries have become increasingly locally owned. It also appears that EO industries generally have higher levels of investment, employment and output growth than inward-oriented industries, reflected in the EO industries' rising contribution to overall manufacturing. However, the government-dominated import-substituting heavy industries experienced considerable productivity improvements in the 1985–90 period with the prospect of lucrative rents available in the protected domestic market. Output/capital and output/labour ratios for non-metal mineral products, iron and steel, and transport equipment, which are dominated by government ventures, improved in this period; however, the aggregated industry data include privately-owned enterprises and hence, should be treated with caution. Also, the apparent strong performance may be due to the government writing off debts on fixed assets through accelerated depreciation allowances, as reflected by the decline in capital-intensities in iron and steel, and transport equipment. Rapidly expanding demand, largely due to the expansion of EO industries and government-sponsored heavy industries, is likely to have raised the performance of inward looking industries. Demand – generated through input requirements, employee incomes and services utilisation – is likely to have strengthened inward looking firms.

Although nominal tariffs on several import-substituting industries were gradually reduced in the 1980s, locally-owned industries were generally still mainly inward looking. Of 12 industries studied (Rasiah, 1996b: Table 9.4), more than half the output demand for 7 was from the domestic economy. Only 5 were outward-oriented, with the electric/electronic and textile/garment sub-sectors being the most export-oriented. Import-substituting industries showed productivity improvements in the 1980s, especially in the 1985–90 period. Falling levels of protection appear to have not negatively affected the performance of import-substituting industries. This could be a consequence of growing domestic demand stimulated by overall economic growth. Given the natural protection offered by geographical conditions and proximity, resource-based industries do not usually require considerable protection to attain competitiveness. As Malaysia is rich in timber, rubber, palm oil, petroleum and tin, it is generally economic to support certain downstream activities when demand reaches sufficient scale. Minimum efficiency scales are also significantly smaller in light and primary heavy industries, which are different from second-order heavy industries. However, cars, iron and steel require significantly larger minimum efficiency scales and greater technological capacity. Of the industries with heavy government involvement, cement has been subjected to price control as well as import quotas. Also, the government-controlled Perak Hanjoong Cement, which manufactures klinker (the main input used in actual cement making), has continued to reap monopoly profits. Perwaja Steel – with accumulated losses of over RM10 billion by 2000 – has been the only government-controlled firm to have performed consistently badly.

The limited success of import substitution forced the switch to export-orientation, beginning symbolically with the Investment Incentives Act in 1968. The promotional efforts of the World Bank, United Nations Industrial Development Organisation (UNIDO) and other international agencies, and the success of other earlier East Asian export-processing zones were important external factors influencing the switch. Meanwhile, local industrial capital began to grow, especially in import-substituting industries, gradually reducing foreign ownership in them. The government has opened free trade zones (FTZs) and licensed manufacturing warehouses (LMWs) since 1972 to ensure better security, co-ordination and control for export-processing activities.

Initially, these export platforms mainly attracted electronics and textile firms as new tax incentives attracted EO firms. Lucrative incentives – such as pioneer status and investment tax credit for periods of between 5 to 10 years – became the main carrots for attracting EO firms. Whenever pioneer status expired, firms were readily given investment tax credits for additional periods of five years. Other firms enjoyed accelerated depreciation allowances. When these expired, some firms opened new plants to enjoy new incentives (see Rasiah, 1993c). In addition, many foreign firms have been allowed to retain total ownership. Hence, though import-substituting industries continued to enjoy high tariffs, financial incentives shifted to EO firms.

EO industries have enjoyed various other government subsidies. Although export incentives that offer double deduction benefits on corporate income tax are given to all exporting firms, given the scale of their exports, the main beneficiaries have been EO industries. Furthermore, EO firms also tend to most utilise the double deduction benefits given for training, as well as for research and development. This is primarily because of rapid international changes in product and process technologies, and the minimum efficiency scale needed to engage in state of the art training, as well as process and design research and development. Apart from resource-based industries (e.g. wood and rubber) and government-controlled car, steel and cement production, foreign firms were the other major beneficiaries of training and R&D incentives. Hence, apart from a few import-substituting industries (especially the government-dominated heavy industries, i.e. steel, cars and cement), EO firms have gradually become more heavily subsidised than import-substituting firms.

It is little wonder then that EO industries have expanded rapidly since the early 1970s. The electric/electronic sub-sector, in particular, has become the most important manufacturing industry in terms of fixed assets, employment, output and exports. However, the average capital-intensities of the electric/electronic and textile/garment industries have fallen since 1985 and 1973 respectively. This is probably due to relatively little new investment from the mid-1970s until the 1980s in textile firms and stronger expansion in the more labour-intensive garment industry after 1985 (see Rasiah, 1993c). The electric/electronic industry experienced a decline in average capital-intensity after 1985 following strong expansion in the more labour-intensive consumer and industrial electronics sub-sector (see Rasiah, 1993a).

Initial expansion in EO industries tended to undermine output-capital ratios, which fell sharply in the 1968–73 period despite falls in capital-labour ratios, a general tendency in that period. Falling capital-intensities were largely due to the growth of export-oriented labour-intensive foreign firms from the early 1970s. It was only from 1979, and especially after 1985, that labour productivity and capital productivity began to improve. Labour productivity rose by an annual average of 3.9 per cent in the 1985–90 period (Rasiah, 1996b: Table 9.6).

Similarly, the operations of foreign transnationals producing for the external market stimulated export expansion. EO industries dominated exports, with the electric/electronic and textile/garment industries together contributing more than 63 per cent of overall manufactured exports (see Rasiah, 1996b: Table 9.7). EO industries also greatly improved their trade balances. However, with the exception of resource-based industries, EO industries had higher import penetration with little trend declines, reflecting fairly weak backward pecuniary linkages. With incentives increasingly oriented to EO industries that faced few ownership constraints, the entire manufacturing sector gradually experienced greater export-orientation.

Since the introduction of the Industrial Master Plan, 1986–95 (IMP) and especially from 1988, efforts to deepen domestic participation and localisation have taken on new dimensions. As shown by Rasiah (1996b: Table 9.4), incentives for exports, training, research and development have been offered (see also Malaysia, 1992). Pioneer status and investment tax allowances are generously offered to 'strategic firms', and since 1991, this has included firms with at least 30 per cent domestic sourcing of inputs. While the government merely offered incentives to EO investments to firms meeting employment, investment and locational targets until the mid-1980s, it has assumed a more pro-active stance since the second half of the 1980s. Incentives have been increasingly tied to technological deepening and to exports, and since 1991, also for increasing domestication of input-sourcing. This strategy has, *inter alia*, encouraged EO transnationals, especially Japanese firms, to relocate their suppliers, including their own subsidiaries, in Malaysia (see Rasiah, 1993a). The policy shift has thus also strengthened backward linkages within the economy. The 1990s have also seen a shift in financial incentives. For example, the government has reduced tax benefits for export-oriented firms other than strategic industries from 100 per cent to 70 per cent.

Especially after 1986, locally dominated inward-looking industries have experienced faster growth, due to a combination of government promotion and growing domestic demand. With the exception of Perwaja Steel, even government-dominated industries have experienced strong growth since the late 1980s. Despite continuing to be primarily inward looking, some import-substituting industries have gradually become outward-oriented. Import-substituting industries that were not set up by the government have gradually been losing tariff protection. Such protection is expected to fall further with the development of the Association of Southeast Asian Nations (ASEAN) Free Trade Area (AFTA) from 1993; tariffs for fast track products are expected to fall to 0–5 per cent by 2000–3. Since AFTA aims to liberalise the ASEAN economies to encourage intra-ASEAN trade, several quantitative and non-quantitative restrictions may also go by the year 2008, which may undermine Malaysia's relatively non-competitive industries and enhance the position of her competitive industries *vis-à-vis* her ASEAN neighbours (see Rasiah, 1994b).

Eng Hardware: from backyard workshop to high precision machine tool factory

Malaysia had a fair number of small-scale metal engineering firms at the time of independence in 1957. The local metal engineering industry, owned primarily by ethnic Chinese, had developed with growing demand from tin mining, infrastructure maintenance, agricultural processing and consumer industries during the colonial period (Rasiah, 1995: chapter 3). The industry was mainly characterised by simple fabrication and foundry work, and operated primarily as backyard workshops. Local firms had little experience in precision engineering works and no automated machinery development capabilities. From simple backyard metal tooling activities, ethnic Chinese firms in Penang began to participate in precision engineering operations in the 1980s. These firms had developed substantial high precision engineering and fully automated machinery manufacturing capabilities by the end of the 1980s. By the mid-1990s, these firms had acquired original equipment manufacturing (OEM) capabilities. Eng Hardware, along with its subsidiary Eng Technology, is one such local concern that has successfully carved out a niche in the high precision machine tool market.

Eng Hardware began as a typical Chinese family venture in 1976. Unlike most Chinese businesses which have developed by servicing the primary and domestic-oriented manufacturing sectors, however, Eng Hardware's growth has been associated with export-oriented manufacturing – particularly semiconductor assembly and test operations in the state of Penang. Eng Hardware's successful growth must therefore be seen in the context of the relocation of mainly American semiconductor firms to Penang. Eng Hardware's development is strongly correlated with growth in machine tool demand by semiconductor firms.

Apart from labour, construction, utilities and some services, semiconductor firms in Penang hardly sourced other inputs from local suppliers in the early 1970s (Rasiah, 1995: chapter 7). When production inputs were acquired locally, these were usually from other foreign firms, e.g. Dynacraft manufactured lead frames. The key production input, fabricated wafers, as well as machinery and components were imported from abroad. Production operations in Penang, nevertheless, encouraged simple metal fabrications involving local firms from the late 1970s. Close collaboration between transnationals and the Penang state government (including the Penang Development Corporation) as well as business networking that facilitated co-operation between selected local firms and transnationals started off the initial metalwork supplier links. However, local sourcing in the 1970s was on a small scale and generally limited to simple fabrication. It was during this time that Eng Hardware emerged as a tooling supplier to semiconductor firms. Its initial participation was limited to simple jigs and fixtures. It subsequently expanded operations to include moulds and dies production.

Advanced Micro Devices (AMD) which began production operations in Malaysia in 1972, first sourced simple off-peak metal fabrication from Loh Kim Teow in 1973. As with other semiconductor firms in Malaysia, AMD had its own in-house workshop which serviced the bulk of such demand, while National Semiconductor had its own machinery subsidiary in Penang, Micro Machining. Eng Hardware's links with AMD, and subsequently, with other transnational firms in Penang, began when the state government began to invite local managing directors to meetings with managing directors of transnational firms located in the FTZs in Penang. In 1978, AMD first sourced some trolleys from Eng Hardware.

Eng Hardware was founded by a traditional Chinese physician, Teh Ah Ba, in 1976. The firm's activities in the 1970s were generally limited to repair work and simple metal fabrications. Demand from semiconductor firms in the 1970s was infrequent and often limited to one-time orders. Eng Hardware's total sales ranged between US$6,800 to US$11,000 annually in the period 1976–8. Little direct technology transfers from transnationals to local firms took place in this period. The parts and equipment locally sourced by semiconductor transnationals in the 1970s did not require high precision engineering. Indeed, Eng Hardware seemed like a typical backyard workshop characteristic of Chinese urban metal tooling works across the country. Traditional artisans who carried on the skills of their fathers or acquired them through apprenticeships typified the skills utilised by Eng Hardware. Eng Hardware's early skilled workers were hired from the urban apprentice market.

Eng Hardware's meteoric rise as a machine tool supplier began following a switch in Intel's machine tooling strategy. Like most other semiconductor firms operating in Malaysia, Intel had used its own in-house workshop for repair work and fabrication. 'Extraordinary' minor fabrications were often subcontracted out to local metal tool firms nearby. Intel began sourcing such services from Eng Hardware in 1979. Intel then started an automation division to enhance its automation efforts in the early 1980s. As in-house activities were increasingly geared towards higher technology aspects of machine fabrication, minor processes were subcontracted out to local firms.

At this time, Lai Pin Yong was appointed Intel's managing director. Intel's move to accelerate the introduction of automation, and the spread of flexible production systems in semiconductor assembly and test operations from the early 1980s accelerated rapid technological obsolescence in the firm, necessitating more frequent fabrication and development. Also, the growing sophistication of assembly and test operations, coupled with continuous shortening of product cycles, expanded the need for proximate metal tool support from simple fabrication to precision machine engineering.

Process flow, factory layout and machinery structures began to experience accelerated transformations from the early 1980s. Intel introduced Just-in-time systems in late 1984; it was the first among non-Japanese semiconductor firms in Malaysia to do so, and one impact was the doubling of productive capability with reduced physical inputs (Rasiah, 1993a). So rapid were the changes that it became uneconomic to import new machinery whenever layouts or production concepts changed. While new machinery continued to be imported, substantial process gains were achieved through constant in-house modifications. Also, the growing need for effective interfacing between machinery users and makers stimulated increased in-house machinery development. While generating substantial production synergies, these developments also created problems for the firm. Machinery production was not only uneconomic (as Intel's own in-house demand was too small to amortise such investments), but the firm could not effectively co-ordinate an entirely new product line.

Interviews with Intel, Hewlett Packard, AMD and Litronix indicate that foreign machinery firms were reluctant to relocate operations in Malaysia. Micro Machining and Texas Instruments in Singapore generally only serviced their own semiconductor subsidiaries. Local sourcing initially appeared impossible as local firms and infrastructure seemed too underdeveloped. Against such a background, market determined prices alone were unlikely to have brought about the development of local supplier networks.

Encouragement by the Penang state government and Lai Pin Yong's appointment as Intel's managing director were instrumental in the development of local supplier networks. Brought up in Penang and enjoying close relations with fellow Chinese in the state, including the ethnic Chinese state government leadership, Lai worked closely with the Penang Development Corporation and quickly forged links between Intel and Eng Hardware as well as Loh Kim Teow. Links were subsequently established with Prodelcon and Metfab – which had been started in 1980 by former engineers of Micro Machining, again strengthened by ethnic ties and political as well as business relationships – which further facilitated effective development of buyer–supplier relationships between Intel and its local supplier firms. Thus, 'trust' helped initiate as well as develop links between Intel and potential suppliers.

Given the technological sophistication and risks involved in manufacturing high precision machine tools, no local firm was initially willing to undertake such operations when Intel first approached them. Local metal working firms had neither the know-how nor the confidence to diversify operations from simple jigs, fixtures, moulds and dies to precision tooling works and automated machinery assembly. With the help of the Chief Minister and officials from the Penang Development

Corporation, Intel managed to convince Eng Hardware to upgrade their operations. Intel was able to offer capital up front to Eng Hardware to enable it to venture into risky precision machinery and tooling operations. Intel also offered know-how, guidelines and prototypes for the manufacture of machinery and components.

In the initial stages, employees from Eng Hardware often visited Intel to acquire production skills. Engineers from Intel also frequented Eng Hardware during this period to ensure more effective production co-ordination as it was in the interest of Intel that Eng Hardware was successfully fostered and developed. After the initial breakthrough with Eng Hardware, it was easier to attract other local firms for fostering. Intel's efforts to foster Loh Kim Teow, Prodelcon and Metfab were aimed at reducing supply bottlenecks and to increase competition among prospective suppliers. Apart from high precision tooling services, Eng Hardware was also manufacturing automated wire bonders and die-attach machines for Intel by the mid-1980s. It should be noted that the changes – primarily in process technology in semiconductor assembly and testing – necessitated by the rapid introduction of more flexible production techniques and automation, as well as product cycle shortening, increased the demand for proximate machine tool activities.

Assisted by process engineers from Intel, Eng Hardware produced its first semi-automated wire bonders in 1981. Starting with technology support from Intel in the early 1980s, Eng Hardware had acquired and developed its own technology to be able to participate on a more equal basis in the development of machine tools by the 1990s. The firm had successfully introduced just in time, statistical process control, total preventive maintenance, quality control circles, time management techniques and professional accounting practices by 1990. As demand rose sharply, Eng Hardware moved its operations from Air Itam to Jelutong in the early 1980s, and then to Bayan Lepas in the late 1980s.

Eng Hardware's machining capability, as measured by the share of computer numeric control (CNC) and automated machines in total machinery, rose from none in 1978 to 12.9 per cent and 30.1 per cent respectively in 1993. Eng Hardware's workforce rose from less than 10 in 1976 to over 200 in 1993. The firm had 4 engineers, 40 qualified technicians and supervisors, and 50 skilled machinists in 1993, compared to none in 1978. By 1993, Eng Hardware was thus able to provide grinding and milling, as well as stamping services with a tolerance level of 0.00005 inch and 0.001 inch respectively.

Sales grew to US$320,000 per annum in the period 1982–3. Output dropped in the period 1985–6 with the downturn in the semiconductor industry. The upswing in the semiconductor industry from 1987 helped raise Eng Hardware's sales to US$6 million in 1990 and US$8 million in 1993. Growing sales and capital deepening helped raise labour productivity, which grew at 30.5 per cent and 12.2 per cent per annum on average in the periods 1986–9 and 1990–3 respectively. Being conditioned by stringent quality and timing standards imposed by transnational purchasers, Eng Hardware began introducing state of the art process technologies quite rapidly. Eng Hardware's technology evolution is shown in Table 5.1.

Exchange rate movements in the second half of the 1980s helped raise Eng Hardware's sales. It should, however, be noted that Eng Hardware's initial rapid

Table 5.1 Eng Hardware: evolution of technology and markets, 1976–2001

Period	*Productive capability*	*Process techniques used*	*Extent of dependency*	*Markets*
1976–79	Manual milling, stamping and grinding of components, jigs and fixtures	Uncodified verbal instructions	Independent use of simple technology	Infrequent sales to varied horizontal markets
1979–81	Manual and electric powered milling, grinding and stamping of components, jigs, fixtures, moulds, dies. Assembly of manual equipment	Codified work procedures	Rising dependence on Intel for technology and markets	Increasing sales to specialized markets, especially to semiconductor firms
1981–87	CNC milling, turning and grinding of components, jigs, moulds and dies. Assembly of semi-automated and automated machinery	Time motion studies, scheduling and accounting methods, quality control circles (QCC)	Strong dependence on Intel for technology and markets. Extensive technology deepening	Around 83% of sales to semiconductor transnationals in Penang. Remainder exported to their subsidiaries abroad
1987–90	CNC precision milling, turning and grinding of components. Assembly of semi-automated and automated machinery. Emergence of Original equipment manufacturing (OEM) status	Just-in-time (JIT), statistical process control (SPC) and QCC	Efforts to break out from dependent relationship with Intel	Sales diversification with exports to disk-drive firms taking around 49% and the remainder going to semiconductor firms
1990–95	OEM status. CNC high precision milling, turning, grinding and stamping of components. Joint-designing of parts and automated machinery with purchaers. Original design manufacturing (ODM) capability but lacks markets for expansion	JIT, SPC, QCC, total preventive management (TPM)	Growing independence from Intel	Sales to disk-drive firms exceed over 60%
1995–01	OEM staus. CNC high precision milling, turning, grinding and stamping of components. Joint-designing of parts with buyers. Manufacturer of actuators and other components for disk drive companies with subsidiaries in Penang, Philippines, China and Thailand	JIT, SPC, QCC, TPM and systems technology	Independent contract manufacturer of disk drive components	Sales to disk-drive firms exceed 90% and the rest to plating equipment users.

increase in sales came in the early 1980s when the critical currencies involved and in which most machinery import invoices were received – the yen, the won, and the Taiwanese, Hong Kong and Singaporean dollars – were fairly stable against the ringgit. As these currencies appreciated against the ringgit following the Plaza Accord and the Malaysian government's devaluation of the ringgit in 1985, Malaysian supplies became relatively cheaper. Currency movements, however, are unlikely to have been the most important explanatory factor in the rapid growth of Eng Hardware's local machine tool sales. Small metal tool firms in the Kelang Valley facing similar transnationals and currency effects rarely enjoyed similar growth in the technological sophistication of their metal tool sales (Rasiah 1994a and b, 1996a). Thus, although exchange rate movements did not trigger off Eng Hardware's expansion, they probably helped enhance growth in the second half of the 1980s.

Like other successful firms, Eng Hardware's continued growth from the late 1980s was not just due to growing demand from Intel. Eng Hardware carefully considered its growing dependence on Intel, which, *inter alia*, controlled the use of technologies transferred from its development department, and also demanded priority for itself over its semiconductor rivals. Machinery, tools and parts designs provided by Intel were not to be used for sales to its rivals.

Despite such constraints, Eng Hardware built on transferred technologies to redesign parts and components for Intel's rivals. Eng Hardware manufactured semi-automated machines and components for AMD, and also supplied precision tool services to Hewlett Packard and Litronix in the second half of the 1980s. Efforts to break out of the dependent relationship with Intel and increased initiatives by Eng Hardware's educated management in the 1980s led to further diversification of its activities in the 1990s (see Table 5.1). Technology transferred to Eng Hardware by transnationals and its own adaptations helped Eng Hardware upgrade its productive capabilities. The 1990s thus saw a gradual decline in the proportion of its sales to Intel.

Eng Hardware, with the support of state government leaders, forged a strong relationship with Maxtor, a disk drive firm located in Singapore. By the end of 1990, 48.8 per cent of sales constituted exports to Singapore. Eng Hardware's supply of disk drive components (including actuators), using just-in-time delivery practices, convinced Maxtor to relocate in Penang. Also of importance were incentives offered by the Malaysian Industrial Development Authority (MIDA) and the exemplary co-ordination role of the Penang Development Corporation (PDC).

The 1990s saw the relocation of other disk drive firms to Penang – Conner Peripherals, Readrite, Komag, Seagate and Quantum. Eng Hardware's main market thus changed from semiconductor firms to disk drive firms in the 1990s. Unlike in the 1980s, however, Eng Hardware has maintained greater independence from the disk drive firms, e.g. by using its own original equipment manufacturing (OEM) technology. It achieved original design manufacturing (ODM) in the 1990s, but has been waiting for capital funding and state support to 'institutionalise' and thus reduce risks associated with expanding such high technology activities.

Extensions to production lines to include critical complementary but dissimilar activities facilitated some amount of in-house machine tool production in Malaysia. Underdeveloped local factor markets meant proximate sourcing options were initially unavailable. The frequency of technological change in semiconductor production, and the uncertainty associated with underdeveloped proximate suppliers made out-sourcing uneconomic. As Coase (1937) and Williamson (1985) have argued, in-house command governance through extensions to the firm hierarchy (internalisation) initially appeared as the most economic solution.

However, as involvement in entirely new product lines required different skills and control structures, semiconductor transnationals increasingly considered out-sourcing. Also, as the volume of machine tool demand generated by semiconductor firms could not achieve scale economies, managements found it undesirable to manufacture all their machine tool input requirements internally. Small and medium sized local firms – with paid up capital of less than RM1 million and employment below 51 during 1979–86, and of less than RM2.5 million and 75 respectively since 1986 – do not require licensing under the Industrial Co-ordination Act of 1975, and thus face less bureaucratic obstacles in running small-scale operations.

The flexibility of small and medium-sized firms, enhanced by the use of multi-functional machinery, has facilitated effective co-ordination involving frequent changes in demand and production specifications. Also, local firms have been able to amortise investments by supplying more firms, which would not be possible for any particular semiconductor transnational competing against the others. It is mainly for this reason that Intel, AMD, Hewlett Packard, Litronix, Motorola, Thomson and International Device Technology – all with subsidiaries in Penang – did not seek regular supplies from National Semiconductor's mature machine tool subsidiary, Micro Machining.

Concluding remarks

Semiconductor transnationals' willingness to out-source their machine tool manufacturing requirements was not initially due to the presence of more economic proximate producers because local firms lacked high precision engineering technology. However, the close rapport Lai Pin Yong and other Sino-Malaysian managers had with the ethnic Chinese-dominated state government officials and local engineering firms brought about greater proximate out-sourcing in Penang. Also of importance has been the organisational and control structure of American firms. Unlike Japanese firms, in which key managerial positions remain controlled by Japanese managers, American firms offered greater autonomy for local managers to make production and sourcing decisions in their subsidiaries in Malaysia. In fact, Intel in Malaysia has been completely run by Malaysians since the 1980s.

Trust has helped strengthen buyer–supplier relationships between semiconductor transnationals and local machine tool firms. Trust has not only compensated for market failure, but also emerged as a key governance component to ensure more effective production co-ordination. Ethnic affinity between the Penang state political leadership and top American semiconductor firm managers, affiliations between local business and political associations and past employment contacts have all been important.

Political circumstances have strengthened ethnic networking in Malaysia so that ethnic-based trust has grown stronger among ethnic Chinese (Khong 1991). Eng Hardware's family management has not included former employees of semiconductor transnationals, but has had access to the semiconductor transnationals' managers through channels organised by the state government and its development corporation. Critical for Eng Hardware's modernisation has been the role of founder Teh Ah Ba's son, Alfred Teh, who qualified as an engineer from Birmingham University in the early 1980s. Alfred Teh has since become a major figure in the state–business co-ordination councils in Penang.

It is clear that semiconductor transnationals – Intel, in particular – have played an important role in the development of Eng Hardware from a simple backyard tooling workshop to a modern high precision engineering factory. The Penang state government and its development corporation have been critical in forging and strengthening links between Eng Hardware and semiconductor firms. Developments in the semiconductor industry favoured out-sourcing from proximate machine tool suppliers. With the underdeveloped factor supply market in Penang in the 1970s, supply arrangements through in-house command governance involving production was initially the best alternative mode.

Rivalry among competing semiconductor firms and the consequent segmentation of markets among individual firms, as well as problems of production co-ordination of dissimilar but complementary products made completely in-house production uneconomic. Intel's decision to foster local suppliers involved considerable trust requiring reciprocity. It was in Intel's interest that Eng Hardware and other supplier firms were developed. Increased out-sourcing by Intel and its direct role in the development of Eng Hardware were governed by a blend of trust, in-house command as well as pecuniary price-cost considerations.

To ensure improvements in the quality and promptness of supplies, Intel, and to a lesser extent, other semiconductor transnationals in Penang consciously transferred state-of-the-art machine tool technology to Eng Hardware and other suppliers. Such transfers, coupled with in-house adaptations and developments, have helped Eng Hardware upgrade its own technological capability. Eng Hardware developed its technology sufficiently to enable it to reduce its dependence on Intel in the 1990s. Indeed, the firm successfully diversified its markets, with disk drive firms becoming its main customers in the 1990s. With the exception of 1985, when a cyclical trough badly affected the semiconductor industry, Eng Hardware achieved double-digit percentage sales growth in every year from 1980 to 1997.

Sapura: rents, technological innovation and competitiveness in a *Bumiputera* Malaysian firm

This case study examines some factors behind Sapura's apparent success in establishing a strong reputation for technological development among private

Malaysian manufacturing firms. Sapura is a relatively young company that has grown in size and scope within a short period of time. Presently, the company is involved in three main business areas – telecommunications, information technology and metal-based industries. The core competence of the firm has been in producing telephone equipment. One of the major factors cited for Sapura's success has been its good connections with key Malaysian government officials. The main objective of this case study is to evaluate the extent to which institutional factors have nurtured this *Bumiputera* (indigenous) Malaysian firm in becoming a successful telephone equipment producer. As Sapura initially depended on foreign technology and joint-venture arrangements, we are also interested in assessing how learning by doing developed from simple reliance on foreign technology. We will also examine the role of entrepreneurial capabilities in company performance.

Sapura started operations in 1975, when the Malaysian economy was growing rapidly. One of the many objectives of Malaysia's New Economic Policy (NEP) introduced in 1970 was to promote *Bumiputera* involvement in business. Shamsuddin Abdul Kadir, the founder of the company, was among the earliest *Bumiputeras* to capitalise on such government policies. Previously an engineer in the Malaysian Telecommunications Department (JTM, corporatised in 1987 as Syarikat Telekom Malaysia, STM), Shamsudddin has had the relevant technical knowledge, experience and contacts in government, particularly in JTM. Like many big businessmen, Shamsuddin is said to have been close to some politicians, notably Prime Minister Mahathir Mohamad. He once served as director of Permodalan Bersatu Berhad (PBB), the holding company of the ruling party UMNO's co-operative, Koperasi Usaha Bersatu (KUB).

All these contacts and experience helped Shamsuddin start his business with a contract to lay cables for JTM in 1975 worth RM2.3 million. With this, he became the first such Malaysian turnkey contractor in Malaysia. This opportunity was almost certainly obtained through his strong government connections. However, the company failed to obtain loans from local financial institutions due to lack of collateral. Sapura would not have fulfilled the tender requirement if not for support from foreign financial sources, particularly through 3M Malaysia.

Sapura has also depended on its foreign partners for much of its subsequent expansion. In 1983, Sapura got a share of the RM2.5 billion cable-laying contract, one of the biggest government jobs before the RM3.4 billion North–South Highway project was awarded in 1987. This contract was divided regionally among four *Bumiputera* contractors – Shamsuddin's Uniphone, Binafon, Electroscon and Sri Communications. Sapura again faced funding problems and needed specialised expertise to handle the contract. Shareholder funds were depleted, and the company plunged into the red because of the huge start-up costs. Again, external financial support helped to pull it through. Sapura brought in two giant Japanese corporations – Sumitomo and Marubeni – as joint-venture partners. The two Japanese companies guaranteed the much needed bank loans amounting to RM70 million.

Sapura has joint ventures with large and established multinationals for most of its other projects. This has given the company the ability to take up projects much larger than its resources might otherwise allow. For example, Sapura is the sole agent for Macintosh personal computers, ancillaries and software, and NEC portable telephones and facsimiles, while Fujitsu is Sapura's joint-venture partner in fibre optics, and Mitsui supplies Sapura with telecommunications equipment for government projects such as earth satellite stations. Sapura has also joined forces with Hewlett-Packard of the USA and Nokia of Finland in other telecommunications activities.

Sapura won a contract from JTM to supply telephone sets during the years 1977 to 1979. Subsequently, the company supplied telephones and payphones for fifteen years under three five-year contracts with JTM. When the second contract ended in 1989, the payphones contract was renewed for another fifteen years. The tender to supply telephone sets, however, was awarded on a two-yearly basis (as a result of the corporatisation of JTM in 1987), and Sapura's contract to supply phones to STM was not renewed. Instead, the contract was given to a Taiwanese company operating from the Prai Free Trade Zone (FTZ) which offered a much lower price for telephone sets. This resulted in criticisms, which caused the government to intervene. When the contract with the Taiwanese supplier finally ended in 1991, Sapura and another local company, Asteria, obtained the contract.

Sapura also operates paging services. These services, which started in Malaysia in 1974, were once a monopoly of JTM, but were liberalised in 1985, with licences issued to *Bumiputera* companies to provide paging services in various localities in Malaysia. Sapura was one such beneficiary. Other important contracts obtained through government connections have been tenders to supply twelve critical component parts for the national car, Proton Saga, since 1990. A contract was awarded to Sapura Machining Corporation to supply two brake parts (brake disc and rear hub), three engine parts (water pump pulley, left and right rocker shaft assemblies), seven transmission parts (reverse shift hug, clutch, release fork shaft assembly, control shaft, stopper body) as well as three shift rail sub-assembly systems. Kyoto Engineering Incorporated, a consortium of six major suppliers to Mitsubishi of Japan, provides technical assistance to Sapura for producing those parts. Sapura obtained this contract – and associated rents – under the local vendor development programme (VDP), launched through and managed by Proton. Thus, Proton is required to share some of its rents (from Proton sales in the heavily protected Malaysian market) with the vendors, mainly *Bumiputera*–Japanese joint ventures.

In general, state connections have benefited Sapura significantly through acquisition of various government tenders and contracts. As a company, Sapura – which began operations with a staff of six in a one-room office in Wisma Central, Kuala Lumpur – has grown in scale and scope. The company has expanded its business activities from telecommunications into information technology, metals-based industries and the automotive parts sector. Telecommunications products and services are the core business of the company, contributing more than 80 per cent of earnings. Table 5.2 shows the importance of telecommunications to Sapura, with its business interests managed by two listed subsidiaries, i.e. Uniphone Telecommunications Berhad (UTB) and Sapura Telecommunications Berhad (STB).

Table 5.2 Main activities of Uniphone and Sapura Telecommunications, 1995 (RM million)

	Uniphone Telecommunications		*Sapura Telecommunications*	
Activities	*Turnover*	*Pre-tax profits*	*Turnover*	*Pre-tax profits*
Manufacturing	113.2	10.4	—	—
Telecommunications	386.0	41.3	408.9	40.3
Investment	1.0	–0.3	—	0.3
Trading	45.0	–0.6	—	—
Property investment	0.9	0.1	2.6	0.6

Sources: Company annual reports.

Telecommunications contributed 71 per cent of turnover and 82 per cent of pre-tax profits to UTB, while the sub-sector was also the major contributor to both revenue and profits for STB in 1995.

Uniphone Telecommunications Berhad was formerly called Malayan Cables Berhad. Sapura acquired a majority interest in the company in 1984, a year after it was awarded the big cable networking contract; Malayan Cables was one of the leading cable manufacturers in the country at that time. In 1988, the company's name was changed to Uniphone Telecommunications Berhad (UTB) after some restructuring. Currently, UTB and its companies are involved in the manufacture of copper rods and communications cables, cable network installation, the manufacture and supply of push-button subscriber phones, and the manufacture, supply and maintenance of public payphones.

Sapura's automobile parts manufacturing has been managed by Sapura Motors Berhad, a publicly-listed subsidiary. Currently, 75 per cent of the company's production is supplied to the national car manufacturer, Proton (Perusahaan Otomobil Nasional Bhd). Its other customers include Perodua, Mercedes-Benz, Volvo, Ford, Suzuki and Mazda. It recently signed a memorandum of understanding (MOU) to supply the Indonesian national car, Timor (*New Straits Times*, 3 January 1997). There are a few subsidiaries which produce manhole covers, mail boxes and cast iron bars.

Economies of scale, growth and profits

These contracts – particularly for telephones, payphones and cable laying – took Sapura into the big league, and have undoubtedly contributed to Sapura's growth and profits. Sapura was the only private company providing telephone sets and payphones in urban areas via long-term contracts with JTM, and then STM. JTM/STM has been the main service operator, providing the core network, while telephone equipment has been supplied by Sapura, which thus gained a virtual monopoly in the telephone and urban payphone markets.

Sapura has enjoyed an additional advantage with telephone sets, because Malaysians are provided with telephones supplied by JTM/STM when they subscribe for a telephone line, unlike subscribers in many other countries who can buy telephones off the shelf. Furthermore, the market for telephones in Malaysia has been large and growing substantially over the years as a result of the greater affluence of the population, increased business activities and other developments in the telecommunications network. The number of telephone subscribers increased from about 0.2 million in 1975 to 2.4 million in 1993, equivalent to an increase in telephone penetration rate (telephone lines per 100 population) from 1 per cent to 13 per cent over the same period (*Bank Bumiputera Berhad Economic Review*, January–March 1995). Although this is lower than the average penetration rate of 49 per cent in developed countries, the country's telephone density is among the highest in developing countries (Rais, 1995). Being the sole supplier of telephone sets to Telekom Malaysia, the rapid expansion of demand for telephones has ensured rapid growth for Sapura. For example, due to increasing demand, production of telephone sets tripled from 134,521 in 1994 to 383,767 units in 1995.

Between 1977 and 1996, Sapura produced four telephone models. The S2000A is the simple push-button subscriber telephone that has mainly been supplied to STM and increasingly to countries like Bangladesh, Mauritius and Papua New Guinea. Other models are more high-tech with more sophisticated features, and are mostly exported to developed countries and sold locally at Sapura outlets (known as *Kedai Sapura*). Production of the S2000HF decreased from 119,858 units in 1994 to 9,702 units in 1995, because as a new model is introduced, production of earlier models falls. The latest model is the S3000. Production of this model has increased substantially from 2,650 units worth RM146,000 in 1994 to 368,123 units worth RM9.7 million in 1995. Meanwhile, total production rose from 134,521 units costing RM6.9 million to 382,767 units worth RM10.3 million.

Payphones in Kuala Lumpur and Petaling Jaya have long been associated with Uniphone, and by 1996, there were 70,000 Sapura payphones (both pre-paid telephone card and coin-operated payphones) operating in Malaysia. Sapura was able to reach such a scale because, as mentioned earlier, a licence was issued in 1977 to Uniphone Telecommunications Bhd (a Sapura subsidiary), to operate payphones in urban areas, while Telekom Malaysia was left to serve the less lucrative rural areas.

Production of payphones has also been very impressive. In 1994 and 1995, Sapura produced between 20,000 to 22,000 units of payphones annually. Payphones' contribution to total telephone manufacturing has also been substantial, worth more than RM100 million in both 1994 and 1995. Thus, payphones have been Uniphone's core business, accounting for about 70 per cent of revenue (*New Straits Times*, 21 February 1996). The substantial share from payphones has been due to its virtual monopoly of the lucrative urban market. Such special privileges have given Sapura economies of scale in production, reflected in increasing turnover over the years. Sapura's sales increased tremendously from RM9.7 million in 1978 to RM162.8 million in 1988 and RM768.7 million in 1995.

Profits of the company have also increased steadily over the years. In the early stage, pre-tax profits of the firm were small, but increasing over the years (Mansor,

1993). In this period, Sapura received two major contracts: RM2.3 million in 1976 to lay cables in the Kuala Lumpur area, and to become a telephone supplier to JTM from 1977. Sapura's growth stage (1983–9) began with the contract worth more than RM600 million in 1983 to lay cables. The Malaysian economy witnessed a short economic recession in the years 1985–6 (Jomo, 1990), during which many companies suffered losses and were forced to retrench employees to cut costs. For Sapura too, this was a turbulent phase. Although turnover suddenly jumped to RM138 million in 1984, and increased rapidly to more than RM400 million in 1985, it suddenly dropped to RM163 million in 1988. The company also suffered losses during some years in this period, particularly in 1984, largely due to the heavy start-up costs for the big cable-laying contract (Mansor, 1993). However, profits began to show in 1988, and increased to RM13 million in 1989. Sapura obtained other local and overseas contracts during this period, including:

- a telephone contract in Bangladesh;
- the contract to install a system for supervisory control and data acquisition (SCADA) for Lembaga Letrik Negara (LLN, the National Electricity Board);
- a licence to manufacture phones in Jordan; and
- a tender to supply telephones in Thailand and Mauritius.

The mature stage began in the year 1990, when turnover rose to RM254 million from around RM200 million in 1989, as pre-tax profits rose to RM18.6 million from RM13.1 million. Pre-tax profits peaked at RM58.8 million in 1994. Turnover, on the other hand, peaked in 1995 at RM768.7 million, though pre-tax profits declined to RM40.6 million. In the 1990s, Sapura was successful in getting many more contracts. In addition, there were many developments in the company's activities such as the launch of flexible card phones, introduction of a new generation of electronic payphones, diversification into the automotive sector, launch of the hands-free voice-activated telephone, etc.

The two listed companies in the group – Sapura Telecommunications Berhad (producing telephones) and Uniphone Telecommunications Berhad (operating payphones) – contributed about 23 per cent and 38 per cent respectively to the group's turnover in 1990 (Mansor, 1993), rising to a total of 86 per cent in 1996. The biggest source of profits in 1992–5 was Uniphone Sdn. Bhd, a company mainly involved in telecommunications-based activities, such as manufacturing modern push-button subscriber phones; manufacture, installation and maintenance of a public payphone network; and installation of fibre optic cables. Teledata Sdn. Bhd was the second largest contributor to Sapura's profits. Manufacture, installation and maintenance of payphones has been the most profitable business for both Uniphone Sdn. Bhd and Teledata Sdn. Bhd. For example, profits from manufacturing payphones accounted for almost 50 per cent of Teledata's total pre-tax profits. This activity has 'saved' the company because there are a few other divisions suffering huge losses.

It is clear that the manufacture, installation and maintenance of payphones and telephone sets have been the major contributor to Sapura's growth. Thus, Sapura's large profits can be mainly attributed to government intervention by limiting and eliminating competition in the payphone and telephone sets markets. The protected domestic market has been crucial for Sapura in strengthening its position in the telecommunications industry in Malaysia. Sapura has made sizeable profits from various contracts it has received via the company chairman's contacts with the relevant government authorities since the mid-1970s.

Rents created and secured through government intervention seem to have been utilised productively by Sapura – more than most other *Bumiputera* rentiers – as evidenced by its heavy investment in R&D to develop technical capabilities and to improve product quality and design. For example, Sapura has succeeded in manufacturing its own locally developed telephone sets by investing in foreign licences and technical assistance, i.e. through learning by doing. It took them eight years to reach this point. Initially, Sapura made telephones under licence from Siemens of Germany from 1980. In 1983, the company obtained a licence to manufacture Bell telephones.

Over the years, Sapura has accumulated considerable experience in manufacturing telephones. To establish itself securely in the telecommunications industry, Sapura sought to manufacture its own telephone from early on. By 1984, with five years of experience in telephone manufacturing under licence, the company had already produced over one million telephone units. Although the company had acquired enough technology and experience to stand on its own feet, it was still constrained from making the required modifications and improvements. For even the simplest circuitry changes, it had to refer to the parent company, and bear all the expenses of related 'expert visits'.

Sapura's R&D efforts have mainly been in product technology, improving products and conducting research for new products in the future, and not in process technology. Sapura mainly does designing and prototyping of in-house products, e.g. telephones and private automatic branch exchanges (PABXs). Of Sapura's input 60 per cent is procured from local sources (*Business Times*, 4 December 1989). The company set up an R&D department in 1984 to make its own telephone. A year later, in 1985, the first home-grown Malaysian telephone, of the S2000 series, was born. Using proprietary technology, the first model was the S2000A, which involved almost RM1 million in development funds. A year later, the first locally manufactured mini-PABX telephone system was launched. This was followed by the second version of Sapura's telephone, the intelligent S2000B, with about RM500,000 spent on its development. This telephone was displayed at the Telecom '87 exhibition in Geneva in October 1987, where it won favourable reviews for its computer-control features, which include a memory bank which stores and recalls within a second over 200 alphabetically-ordered numbers, and other features such as automatic dialling and call barring. This telephone has been exported to Japan, the USA, Germany and many other countries. The telephone also won first prize in the utility innovation section in the Malaysian Invention and Design Exhibition. Meanwhile, the company also successfully developed its own direct paging software programme. Sapura has increased its budget for R&D over the years, and also aims to produce its own brand of mobile telephones.

In order to improve capacity to face future challenges, and to further emphasise its role, the board of directors of Sapura Holdings agreed to incorporate the R&D department as a corporate subsidiary. Hence, Sapura Research Sdn. Bhd was formed in February 1991, with a paid-up capital of RM2 million. The rationale for setting up the company was also to gain greater operational autonomy. Shortly after its incorporation, Sapura Research announced another achievement – development of a hands-free, voice-activated telephone, the S2000HF, with home-grown technology. In the first year of operations, the R&D unit was allocated RM2.8 million, with the amount increasing steadily to about RM5–6 million in 1990/1. This amounted to about 1 per cent of the group's turnover during those years. In 1992, R&D expenditure was estimated to be around RM10 million, which was about 2 per cent of turnover. However, due to the diversified nature of the company, measuring R&D expenses against turnover may be misleading (Mansor, 1993). The proportion of R&D expenditure against telephone sales has been about 10 per cent – quite comparable with established multinational companies like IBM, Matsushita, Philips, Xerox and Ericsson, for whom the proportion ranges from 8 to 15 per cent.

Only printed circuit boards (PCBs) and chassis/mould are produced in-house (boards are bought from local vendors, while the surface-mounting is done by Sapura). Metals, plastic, as well as some of the chassis and moulds are supplied by local manufacturers. Other inputs – such as test instruments, metals, plastics and IC chips – are imported, with the most important import being integrated circuits (ICs)/specialised chips. This is because although there are many semiconductor manufacturers in Malaysia, most of their production is for export. In addition, Malaysian companies are still producing application-specific ICs.

The telecommunications sector is one area where the development and application of new technologies is very active. Progress in new electronic-based technologies during the 1980s has made large MNCs adopt integrated computer-aided design (CAD) and manufacturing (CAM) systems. While this permits new products to be commercialised very quickly once they have been designed – i.e. shortening the lead-time from the product conception stage to the design stage – it also allows manufacturers to respond rapidly and flexibly to customers' specifications and tastes. Consequently, there is a perceptible trend for manufacturing enterprises to emphasise product diversification, design, and servicing of their products, apart from production itself (Anuwar Ali, 1992: 66).

Multinational companies – such as Siemens, Thomson, Hitachi, NEC, Ericsson, Toshiba and Philips – obviously dominate industrial R&D and technological innovation in the international telecommunications industry. These giant corporations set the pace and direction of R&D, and thus define the technology frontier. Catching up with these giants, particularly in technology, is not easy for a latecomer like Sapura. Furthermore, being still in a catching-up phase, technological advancement in Sapura seems to be incremental, gradual and achievable through many small modifications, rather than being based on major breakthroughs. In addition, products or processes developed through Sapura's R&D efforts that are regarded as new for Sapura may not be very new in the world.

Apparently, Sapura's R&D efforts have not helped to reduce production costs by very much, and have thus not helped much to improve international competitiveness. For example, in 1989, a Taiwanese company, Formula Electronics was selling telephone sets for RM37.20 a set, compared to Sapura's RM54 per set. This meant that Sapura's price was 45 per cent higher than that of the Taiwanese supplier. Sapura's officials admit that it has not achieved international competitiveness in terms of price, and that Korean and Taiwanese producers are generally much more competitive in the world market.

Technological learning also requires the industrial work-force to be well equipped to acquire technical skills, while shop-floor technicians, engineers and technically trained managers are required in increasing numbers. This means that secondary and technical education has become more important. However, most of the workers in Sapura are not highly educated, though they seem to be generally better technically trained compared to Taiwanese assemblers in Ipoh and Perai. Of Sapura's factory workers only 85 per cent have upper secondary, lower secondary or elementary school qualifications, while 15 per cent of the employees have university and technical institute training.

Sapura is thus essentially an assembly-type production company, which largely employs cheap and unskilled workers with low levels of education, which in turn limits the capacity for rapid technological change. Given its employment structure and education level of its employees, Sapura is obviously far behind the standard and quality achieved by its competitors in East Asia. Therefore, it is questionable whether Sapura will ever be able to compete successfully with its competitors in the world market in terms of price and quality.

Rapid progress in new electronic-based technologies in the industry makes the catching-up process even more difficult for Sapura. But catching up with the world technological leaders may not be the main priority for Sapura. This is because it relies mainly on protected markets and hence, export sales are not vulnerable to price competitiveness. Furthermore, Sapura mainly sells the simple push-button model to developing countries, because the tenders it has received so far have been from the government-owned Telecommunications Department, where cheap and easy-to-use telephone sets for the general public are preferred.

Sapura's high costs of production and slow technological change have made exports to developed countries difficult, with sales to these countries negligible. It exports sophisticated top-of-the-line telephones, such as the S2000HF and S3000 models, to Japan, the USA, Canada, Germany, Austria, France and Belgium. The American market has been one of the most difficult to penetrate. In 1989, Sapura appointed a California telecommunications-based company, Landsperger and Associates, as its representative to co-ordinate its North American sales and marketing. The arrangement, however, did not materialise, and Sapura tried to have another such arrangement in 1991, but without much success. Foreign sales have mainly been to other industrialised countries such as Singapore, the Netherlands, Germany, Iceland and Japan.

Conclusion

Sapura's success can be traced, in large part, to successful medium-term identification and pursuit of lucrative engineering-based business opportunities, initially in telecommunications and, more recently, in the automotive industry. Sapura has gone through various challenging stages of development, in which it has managed to acquire and build new capacities from various favours (rents), e.g. the relevant business licences and contracts which have enabled it to develop resources and capabilities to become more profitable.

Sapura's R&D effort is commendable, especially considering its size and experience compared to bigger local and multinational companies. However, Sapura's R&D efforts have not yet enabled the company to become truly competitive internationally in terms of price. Therefore, for example, Sapura phones still cannot compete with its competitors from Taiwan and Korea, who are the least cost telephone producers in the world. The high costs of production mainly stem from its dependency on foreign technology and imported inputs. Even though Sapura exports its products to foreign countries, many such sales were not attained due to cost competitiveness, but through the company's partnership with transnational giants and Malaysian government influence abroad.

However, the possibility of Sapura achieving international competitiveness through its own technological advancement seems remote as the company's R&D efforts largely involve upgrading its main products, i.e. telephones, software, etc. which face stiff competition in the world market. In addition, the company also lacks skilled and trained staff to keep up with the fast changing international telecommunications technology. Realising this, Sapura concentrated on getting contracts in developing countries, where there is extensive government intervention in the procurement policies of big local companies, particularly in the telecommunications sectors.

The sale of 75 per cent shares of Uniphone Sdn. Bhd and Sapura Digital Sdn. Bhd – two important Sapura subsidiaries making substantial profits – indicated the company's shift of emphasis to another even more lucrative sub-sector offering higher rents, i.e. automotive parts, as rents in Malaysian telecommunications have been eroded by various changes in the industry. The automotive industry is still highly protected, and there are significant rents to be captured with the vendor development programme. There also appear to be some lucrative opportunities to be tapped in multimedia information technology (IT), with the launching of the Multimedia Super Corridor (MSC) by the Malaysian Prime Minister, though it is also feared the big foreign transnationals will capture most of the special privileges being offered.

Proton: building a national car industry in Malaysia

This section analyses the experience of the Malaysian National Automobile Enterprise, or Perusahaan Otomobil Nasional Berhad (Proton), for which state-created rents have been instrumental. An attempt is made to examine the types of rents enjoyed by Proton, the mechanisms used to manage them, and their consequences.

Background

The Malaysian government's heavy industry programme began in 1980 with the incorporation of the Heavy Industry Corporation of Malaysia (HICOM). Mahathir Mohamad was then Deputy Prime Minister as well as Trade and Industry Minister under Hussein Onn's premiership. HICOM was given national responsibility for heavy industry development. HICOM's subsidiaries have sought to utilise rents to catch up internationally, but also to support *Bumiputera* industrial progress. HICOM programmes received a major boost with Mahathir's appointment as prime minister in 1981. The first half of the 1980s saw the commencement of cement, steel and motorcar production by HICOM subsidiaries – Perwaja Terengganu, Kedah Cement and Proton respectively.

HICOM has also pursued a key objective of the government – the creation and expansion of *Bumiputera* manufacturing enterprises manned by predominantly *Bumiputera* technical and managerial personnel. Until recently, *Bumiputera* participation in manufacturing had been largely limited to bottom-rung operator activities. The only managerial positions they dominated in transnational firms operating in Malaysia were those for personnel, human resources, public and government relations (Rasiah 1993a: chapter 5). The preference for such *Bumiputera* managers has generally been due to the work-forces being dominated by *Bumiputeras* and their presumed better ability to deal with *Bumiputeras* who dominate the public sector. They have generally not been appointed to positions key to the firms' production activities.

Proton was incorporated in 1983, and the first cars rolled out of the plant in Shah Alam in 1985. It was set up as a joint venture involving Malaysian state capital and Japanese private capital.[5] HICOM remained the principal shareholder in 1995 (Table 5.3), when the state effectively controlled 54.5 per cent of its shares while Mitsubishi owned 17.2 per cent, down from its 30 per cent share at the outset. The firm was listed in the Kuala Lumpur Stock Exchange (KLSE) on 26 March 1992. Initial production was entirely for the domestic market. Exports started in 1986, with 25 cars shipped to Bangladesh. The UK soon became the biggest foreign market with exports reaching 17,440 in 1993 (Proton, 1995: 7). By 1995, Proton had exported to 28 countries. Proton had operating losses in the period 1986–8, after which it made profits every year, reaching RM407.9 million in 1992 (Proton, 1995: 10). Its rapid expansion led to the firm doubling its annual production capacity from 80,000 cars in 1985 to 160,000 cars in 1995 – making it the largest automobile car plant in Southeast Asia. The 1997–8 financial crisis affected sales, which fell in 1998 to 52 per cent of the 1997 figure. Sales have rebounded following the easy credit for car purchases policy implemented after the introduction of capital controls in September 1998. The late 2000 deferment to 2005 of the ASEAN Free Trade Agreement (AFTA) dateline for reducing barriers to automotive imports will extend protection of the domestic market and related rents for the domestic car industries, thus ensuring an additional lease of life.

Proton has become an important platform for generating new *Bumiputera* technical and managerial expertise. Employment has grown strongly over the ten

Table 5.3 Proton: share-ownership structure, September 1995

Shareholders	*Share (%)*
HICOM Holdings	27.5
Khazanah Holdings	17.5
Mitsubishi	8.6
Mitsubishi Motor Corporation	8.6
Government agencies	9.5
Other local and foreign investors	28.3

Source: Proton (1995: 1).

Table 5.4 Proton: work-force structure, 1989–95

	1989	*1990*	*1991*	*1992*	*1993*	*1994*	*1995**
Executives	242	259	407	480	634	667	725
Indirect staff	784	928	1019	1075	1325	1463	1600
Direct staff	689	1322	1465	1586	1997	2113	2518
Total	1715	2509	2891	3141	3956	4243	4843

Source: Proton (1995: 11).
Note: * September 1995 figures.

years (see Table 5.4). *Bumiputeras* accounted for over 98 per cent of all employees in the factory. Largely employed without prior experience, the technical work-force has been trained in Japan and in-house. With considerable rents from high tariffs and excise duties, Proton appears to be a successful state-sponsored manufacturing venture. The firm had five subsidiaries and eight associate firms in 1995. Subsidiary firms were owned either directly by Proton or through Edaran Otomobil Nasional (EON) or HICOM. The share of *Bumiputera* employment in these firms resembles Proton's. Associate firms have been engaged in component manufacture and car assembly, with substantial equity shares held by other firms. Among the associated firms are joint ventures involved in assembly in the Philippines and Vietnam which use some *Bumiputera* managerial expertise.

The government launched a localisation programme to complement Proton's production plans. Three reasons to explain this are plausible. First, to counter the paralysing effect of rising import costs from Japan following the Plaza Accord of 1985 by increasing local value-added so as to reduce foreign exchange losses and external dependence. Second, to meet the Generalised System of Preferences (GSP) domestic content requirements. Third, to promote *Bumiputera* enterprises through supply contracts for Proton. Through its vendor development programme, Proton had created 138 domestic vendors by 1995. By 1995 too, Proton had managed to source 3,511 components domestically – 394 in-house, 3,076 from domestic vendors and 41 'resourced', i.e. from domestic vendors importing from abroad.

Using GSP criteria, Proton had achieved 67 per cent domestic content in 1995 (Proton, 1995: 9). By its own local material content policy (LMCP) criteria, Proton had 80 per cent local content in 1995.

This is a remarkable achievement given that Proton only started operations in 1985. South Korea – which had a localisation ratio of 21 per cent in 1966–9 – achieved 92 per cent domestic content in 1981–4 (Doner, 1991: Table 1). Unlike South Korea, which has no ethnic promotion policy, Proton's vendor programmes have involved the development of (previously inexperienced) *Bumiputera* vendors, and hence, the localisation process – like Proton's own development – has involved catching up with both national and international technology frontiers. The effectiveness of this programme is discussed in the next section.

Creation and management of rents

Assessing Proton's performance is not an easy task. By the end of 1995, Perwaja had accumulated debts amounting to RM2.9 billion – a figure that does not include debts accumulated prior to 1988, which were written off following the appointment of Tan Sri Eric Chia as managing director. Proton has continued to generate exports, although exports' share of total production peaked in 1992 (see Table 5.5). By most accounts, the firm has done well financially in the 1990s. Protection – through the use of tariffs and excise duties – has helped shelter it from foreign competitors including foreign-controlled producers assembling domestically. Government policy has ensured adequate funding for Proton's undertakings, especially in the initial period when its viability was uncertain. Its financial performance after 1989 has ensured subsequent private funding. In fact, when the firm was listed on the KLSE in 1992, it was heavily oversubscribed.

Since the long-term viability of the industry will depend on the eventual reduction if not elimination of government-created rents so that they are only

Table 5.5 Proton: sales and exports, 1985–95

	Domestic ('000) (1)	*Exports ('000) (2)*	*Total ('000) (3)*	*2/3 %*	*% of exports to developed markets*[a]
1985	7.5	0.0	7.5	0.0	0.0
1986	24.1	0.0	24.1	1.0	0.0
1987	24.9	0.4	25.3	1.6	37.7
1988	42.5	0.9	43.4	2.1	73.3
1989	52.7	11.9	64.6	18.4	98.5
1990	72.5	13.1	85.6	15.3	98.5
1991	84.8	15.1	99.9	15.1	98.2
1992	80.4	18.8	99.2	19.0	98.2
1993	94.1	20.3	114.4	17.7	97.1
1994	111.3	15.0	126.3	11.9	91.5
1995[b]	144.0	21.1	165.1	12.8	88.7

Source: Computed from Proton (1995).
Notes: [a] includes Singapore; [b] forecasts.

limited to stimulating innovative activities, e.g. subsidies for R&D and training, the effective deployment of rents in the interim becomes critical. However, the World Trade Organisation (WTO), Asia Pacific Economic Co-operation (APEC) forum and AFTA all require trade liberalisation, that would remove much if not all such rents, which would jeopardise the viability of the industry. Thus, this section examines government support for Proton, and the extent to which monitoring, appraisal and action have been taken to reduce rents. One cannot rigorously estimate likely returns on investment in the absence of protection due to lack of information on Proton's finances and subsidies enjoyed by Proton.

Infant industry supporters have typically recommended protection and subsidisation – with protection preferred over subsidisation due to limited capital resources in developing economies – so that import prices exceed the infant industry's domestic prices (Lewis, 1955; Myrdal, 1957; Kaldor, 1979). Unlike Prebisch (1962), who recommended import-substitution to overcome balance of payments problems, infant industry proponents actually saw the development of productive capacity as being more critical. In their model, tariffs should be lowered gradually as the infant industry matures.

As shown in Figure 5.1, tariffs and subsidies should be lowered as the long-run average cost (LRAC) falls. It is generally only possible if the domestic market is sufficiently big to achieve minimum scale efficiency (MSE), with disciplinary mechanisms introduced to force efficiency improvements in the absence of competition. Without dynamic efficiency improvements, the infant industry will only move down the $LRAC_1$ path, with MSE reached at quantity Q_1 at unit cost C_1 – which is higher than import price P_1. Such static efficiency gains will not

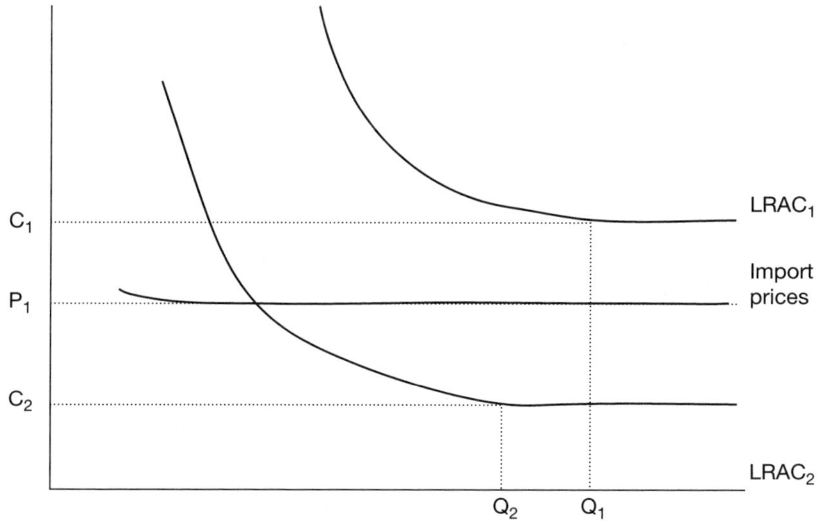

Figure 5.1 The case for infant industry protection

make the infant industry internationally competitive. The LRAC will only shift downwards to $LRAC_2$ if the infant industry achieves real efficiency gains through rapid technical change. Effective management of rents arising from tariffs and subsidies should enable the firm to achieve its MSE level at Q_2 – at which point the firm does not require rents any more to be competitive, as the $LRAC_2$ at point C_2 is below the import price P_1. At this point, the firm is considered to have matured and will be able to offer domestic consumers a price lower than the import price. Consumer welfare losses in the period until quantity Q_2 can be justified on the grounds of saving foreign exchange and creating a viable local industry that could eventually achieve dynamic efficiency gains.

If investments in R&D generate technical change that improve productive efficiency, then the LRAC would move downwards from $LRAC_1$ to $LRAC_2$ so that long-run costs at point Q_2 – i.e. the new MSE quantity – will fall to C_2 (see Figure 5.2). In other words, a share of the monopoly rents enjoyed in the domestic market should be invested to help Proton improve its competitiveness and to help it catch up with and overtake at least some of its international competition.

Through export targets, gradual withdrawal of rents, constant appraisal of real technical gains and state–business co-ordination, Japan, Korea and Taiwan managed to develop firms that successfully moved down the $LRAC_2$ path (Figure 5.1), and eventually overtook many international incumbents by achieving real efficiency gains (Amsden, 1989; Shinohara, 1982; Fransman, 1986; Wade, 1990). Otherwise, the infant industry will continue to enjoy rents without striving for efficiency improvements, such that the LRAC curve moves down very slowly, tracking a much gentler gradient as in $LRAC_1$. Not only will it require a larger quantity to reach MSE, $LRAC_1$ will also remain higher than $LRAC_2$. As noted

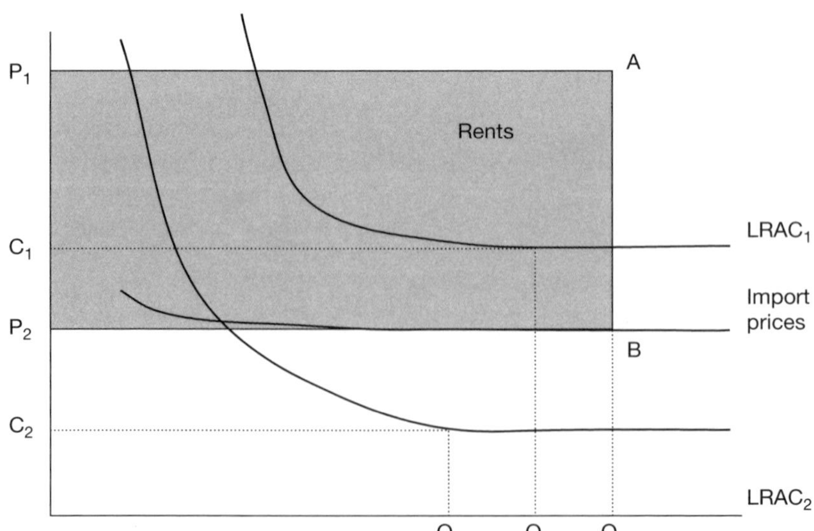

Figure 5.2 Rents and technical improvements

earlier, at quantity Q_1, $LRAC_1$ lies at C_1 – which is higher than P_1 and C_2. This is what happened to several of India's import-substitution industries.

The question to ask now is whether Proton has been tracking the $LRAC_2$ path, as achieved by the successful Northeast Asian firms. This is a difficult question to answer. Unlike the Northeast Asian experience, for which information has been made available for more rigorous assessment – which helps the state obtain alternative viewpoints cheaply – most data on state-sponsored enterprises in Malaysia are kept secret. This section attempts to examine rent management in Proton, using less exhaustive methods.

Proton was established by the state because private *Bumiputera* capitalists were not in a position to initiate it. They had neither the capital nor the technical know-how – including assembly experience – to produce cars. Some non-*Bumiputera* Malaysians had some experience in car assembly and entrepreneurial experience in other modern manufacturing activities. Non-*Bumiputera* capital would also have required substantial state-created rents to have had any chance of succeeding. However, since Proton was not just aimed at raising Malaysian capability but, more specifically, *Bumiputera* manufacturing capability, the option of supporting potentially more capable non-*Bumiputera* entrepreneurs did not arise and probably would have been politically unfeasible. Given the lack of capable *Bumiputera* entrepreneurs in the early 1980s, especially in manufacturing, the question of holding competitive auctions did not arise.

The earlier success of Japan and Korea seems to have convinced Prime Minister Mahathir that catching up with the international technology frontier was both necessary and possible. Hence, there was little effort to examine Proton's likely viability before its launch (Jomo, 1985; Chee, 1985; Khor 1987). In South Korea, as in Malaysia, widespread criticisms had been directed at the government's heavy industry ventures (see also Lal, 1983), including the rejection of loan applications by the World Bank (Amsden, 1989; Chang, 1994). The South Korean state nonetheless went ahead with the programme, accumulating a huge foreign debt in the process. By the mid-1980s, it had become clear that the South Korean government had succeeded in making a success of its heavy industrialisation effort. The South Korean government had one less disadvantage compared to the Malaysian state because it was committed to advancing an 'entrepreneurially underdeveloped' ethnic group.

The extent of public–private co-ordination at the time of Proton's launch was minimal. Proton's first directors had virtually no experience or links with activities related to car manufacture. There is also little evidence of local automobile assembly managers before the mid-1980s being involved in Proton, perhaps due to the government's desire to 'by-pass' the existing Chinese-dominated car assembly industry (Jomo, 1986; Jesudason, 1989; Rasiah, 1996b). Nevertheless, Proton's marketing and servicing activities in Malaysia – handled via Edaran Otomobil Nasional (EON) – have attracted and involved substantial participation by Chinese businesspeople. In fact, Chinese have continued to dominate the marketing of the car in Malaysia as substantial rents were to be gained from marketing the (protected) car. For the government, the sales and repair services of ethnic Chinese have been critical for gaining and sustaining customer confidence. Unlike with marketing and servicing, however, the government has relied little on Chinese expertise in manufacturing the car.

Both subsidies and protection have characterised the growth of Proton. Preferential loans guaranteed by government have ensured that Proton has enjoyed interest rates lower than available market rates and greater funding support than would otherwise be normally available (see Landau *et al.*, 1990; Machado, 1994). It is unclear if these preferential rates have continued. Officials from Proton say that they no longer enjoy corporate tax rebates. However, like other export-oriented firms, Proton has continued to enjoy the export credit refinancing allowance – which is a subsidised loan given for exports. The firm has also enjoyed a double deduction allowance for R&D expenditure. The reported RM82 million spent on R&D in 1992, for example, would have exempted an additional RM82 million from profit taxes. The firm also enjoyed a double deduction allowance for training expenses from 1988 to 1992, but this figure was reported to have been small, e.g. only around RM5 million in 1992. It should be noted that the export, training and R&D subsidies enjoyed by Proton have also been enjoyed by other firms, including foreign transnationals, which account for most of the double deduction training allowance claims (Rasiah and Osman-Rani, 1995). Besides, incentives for R&D have been necessary, even for firms at the technology frontier, to stimulate such investments.

The nominal rate of protection on automobiles varied from 150 to 310 per cent in the period 1985–93 (Chan, 1994). Although tariffs and excise duties have been highest on completely built-up (CBU) units, components have also come under relatively high duties due to a deliberate policy to develop domestic suppliers. Hence, the effective rate of protection for automobiles in 1990 was 52.6 per cent (Broadway *et al.*, 1991). Given the subsidies Proton suppliers enjoy through preferential funding, the net subsidy equivalent of Proton is likely to be much higher than the previous aggregate for the automobile assembly industry. Interviews suggest that there has not been any significant reduction in the tariffs protecting Proton. Hence, prices facing domestic consumers have remained significantly higher than world prices. Using Figure 5.2 for illustrative purposes, it can be seen that domestic consumers have been facing price P_1 when compared to price P_2 enjoyed by foreign consumers – allowing the firm to enjoy rents from domestic sales of quantity Q_3, equivalent to the shaded rectangle P_1ABP_2.

Domestic Proton car prices are still higher than Proton car prices in export markets. Interviews suggest that the car's marginal cost exceeded foreign prices until the 1990s. A manager was quick to add that this was also the case with domestic prices until around 1988 – which is likely, given the losses incurred by Proton until 1989. Proton has made significantly lower profits for every car sold in foreign markets while continuing to reap handsome rents from domestic consumers by continuing to enjoy high protection. In the newer external markets, small sales volumes have not even allowed Proton to break even.

The huge rents made from domestic consumers, and the decline in marginal costs due to scale economies has enabled Proton to avoid losses in foreign markets and even made some of them profitable. With such rents enjoyed by Proton, direct workers earned wages around RM1,500–2,000 a month with overtime in 1995. They also enjoyed other benefits, such as discounted prices on Proton cars, once every two years. It can also be argued that domestic consumers have indirectly subsidised exports as they have paid high prices before the MSE level was achieved. Such rents are expected to continue in the next few years as there is, as yet, no sign of an imminent decline in protection.

The government has aggressively promoted the Proton car in some foreign markets. In fact, bilateral official trade negotiations have been used to promote Proton exports. Relentless promotion – especially by the Prime Minister – has been instrumental in raising export sales. If the export share is taken as a measure of competitiveness, the firm has made progress. The share of exports in total sales rose from nil in 1985 to a peak of 19.0 per cent in 1992, before falling to 11.9 and 12.8 per cent in 1994 and 1995 (see Table 5.5). Exports to developed economies have dominated foreign sales, accounting for over 98 per cent in the years 1989–92. However, since foreign sales have not generated significant profits, this measure is not a good indicator for assessing Proton's competitiveness. It only shows that foreign customers will purchase Proton cars if priced low enough. However, since indirect export subsidies have also been important in the growth of heavy industries in Japan and Korea, it would be premature to classify Proton as a failure.

The government's launching of a second automobile firm, Perusahaan Otomobil Kedua (Perodua) suggests some willingness to open the Malaysian market for some competition, albeit from another state-sponsored firm. Also, the Kancil produced by Perodua is smaller and does not compete directly at the same engine capacity and vehicle size level with cars manufactured by Proton. Nevertheless, it has probably cut into potential sales by Proton. Proton's share of the domestic market fell slightly from a peak of 74 per cent in 1993 to 73 and 72 per cent in 1994 and 1995 respectively (Proton, 1995: 6).

As noted earlier, Proton's performance should eventually be judged in terms of the gains generated for the national economy against the welfare loss borne by Malaysian consumers. The 15–20 year gap behind the technology frontier, does not suggest rapid progress towards the technology frontier. The acquisition of Lotus in 1996 may have given Proton new technological capabilities to accelerate progress up the technology ladder, but the actual gains of the acquisition have still not been measured; and dissimilarities between the products of Lotus and Proton may actually limit such gains. It is clear that protection for the car industry has yet to decline despite huge profits reported by Proton. Why has the government not begun reducing tariffs, especially since the MSE – based on current technological capability – has been achieved?

Four plausible explanations can be suggested. First, the contract between the government and Mitsubishi could have ensured that the latter would be the prime beneficiary of rents – only leaving a relatively small share for the government.[6] Second, as the more sophisticated infant industry advocates may argue, to achieve real efficiency gains, the firm requires substantial R&D investments to help accelerate progress towards the technology frontier. Proton was reported to have invested RM82 million in R&D in 1992. However, an engineer from Proton privately conceded that the firm was around 15–20 years behind the technology frontier firms in 1995.

Also, Proton has yet to develop independent engine manufacturing capability. It is still dependent on its Mitsubishi partner for the engine and the gear-box, and for expertise in several other critical aspects of car manufacture. In fact, Machado (1994) has argued persuasively that Mitsubishi managers' participation has been critical for upgrading and sustaining the high quality standards required in developed markets, suggesting Proton's heavy technological dependence on the Japanese partner. Table 5.6 shows several technology transfer agreements (TTAs) still tying Proton to foreign technology. Indeed, the car is supposed to embody 'Japanese technology, Malaysian style' – as advertised in England.

Finally, a significant share of the costs incurred by Proton still goes to its technology suppliers. In Japan and South Korea, pro-active and visionary governance – by *ex ante* vetting, monitoring technology transfers, *ex post* appraisal and development of institutional capabilities to raise absorptive capacities – have facilitated extensive technology diffusion (Johnson, 1982; Rasiah, 1995; Lall, 1996). In Malaysia, the government has lacked technically and economically proficient and politically shrewd technocrats and bureaucrats to maximise gains for local licensees when dealing with foreign licensers and other technology suppliers. Technology transfer agreements (TTAs) in Malaysia have not involved any *ex post* monitoring and appraisal, while the *ex ante* screening is poorly handled (see Rasiah, 1996b).

Utilisation of parts from foreign transnationals in Malaysia (e.g. Robert Bosch and Nippon Denso supply Blaupunkt stereo sets and air conditioners respectively to Proton) and efforts to diversify component sourcing to reduce dependence on Mitsubishi (especially with the rising value of the yen) led to serious government efforts to broaden the sources of TTA partners. Indeed, as with most major directives involving Proton, Mahathir himself has been pushing the diversification programme. Officials from the Economic Planning Unit and Proton note that

Table 5.6 Proton: technology transfer agreements, 1995*

Country	*Joint-venture*	*Technical assistance*	*Purchase agreement*	*Total*
Japan	16	35	4	55
Germany	3	0	0	3
Taiwan	5	1	0	6
Korea	6	0	0	6
Australia	2	1	0	3
Others	3	3	0	6
Total	35	40	4	79

Source: Proton (1995: 9).
Note: * as of September.

Mahathir has been key to the co-ordination of Proton's activities. Despite its utility in gaining priority for Proton as well as public and private support in Malaysia and abroad, excessive dependence on a busy prime minister has also adversely affected Proton's development.

The vendor development programme (VDP) launched through Proton works through an umbrella framework where captive rents from Proton sales are shared with the mainly *Bumiputera* vendors and their partners. Since most of the *Bumiputera* vendors are recent start-ups with limited business and manufacturing experience, Proton has had to pay high prices to these suppliers (see also Rasiah, 1996b). Proton has to source parts and components from high-cost government-designated producers in the hope that, in the long run, they will improve efficiency and reduce prices below those of imports, thereby enhancing Proton's competitiveness. Part of the monopoly rents enjoyed by Proton as shown in Figure 5.2 actually goes towards the development of domestic suppliers. If such extension of the domestic value-added chain helps lower input prices so that they become cheaper than imports, then it will help push down LRAC from $LRAC_1$ to $LRAC_2$.

However, Proton's production officials say that, while they have succeeded in increasing local content in a short time to 80 per cent, the other consequences have been mixed. Most suppliers have actually raised costs to Proton as they have not managed to significantly improve efficiency. The main benefits generated so far have been improvements in delivery time and some cost reduction for certain components. Also, the suppliers have themselves been importing considerably, meaning that much of the 80 per cent local context has actually been imported. In addition, interviews suggest that a significant share of the rents accruing to suppliers have been appropriated by Japanese suppliers attracted to Malaysia by Mitsubishi – which, despite sharing equity with *Bumiputera* partners, has retained control over the key technologies used, besides securing the lion's share of the rents thus obtained.[7]

Furthermore, the government has no rigorous mechanisms to monitor and improve performance in order to adjust tariffs downwards to reflect new levels of efficiency achieved by Proton and its suppliers. Given the monopolistic privileges it has enjoyed, it is unlikely that Proton will initiate such progressive tariff reductions. Interviews support the view that the bureaucracy has little relevant technical competence, let alone inclination to assess and improve Proton's performance and to adjust optimal tariff levels to sustain growth, improve efficiency and enhance welfare. Yet, performance auditing can easily be done. Profits and internal rates of returns can easily be computed for different tariff scenarios taking account of both outputs and inputs. Further assumptions can be made based on informed estimates of likely domestic demand at different tariff levels.

It is unlikely that Proton can compete with foreign cars without significant tariff protection for quite some time to come. In September 1995, customers in the United States, Australia and Malaysia paid RM45,000, RM50,000 and RM115,000 respectively for a Toyota Camry. The prices of the same car in Australia and Malaysia in April 2000 were RM62,500 and RM135,000 respectively. If the Toyota

Camry, or a cheaper model, such as the Toyota Corolla or Nissan Sentra, were available at US tariff rates, it is difficult to imagine significant numbers of Malaysian consumers still wanting to buy the Proton. Rigorous performance evaluation can be used to force Proton to improve efficiency, and to facilitate a gradual reduction in tariffs to 0–5 per cent by 2005 – the new extended AFTA dateline for automobile imports deregulation for Malaysia.

High rents have enabled Proton's suppliers to enjoy lucrative returns without being internationally competitive. Due to the lack of information, one can only conjecture about the significance of the problem. If it can be established (e.g. by using simulations of performance when tariffs are removed) that Proton is not likely to be internationally competitive nor capable of generating efficiency gains, it could then be argued that the venture itself has only enriched Proton and others linked with it, including Mitsubishi and the vendors. There is some evidence of politically favoured firms having access to Proton's rents (Jomo, 1985; Khor, 1987), but this in itself does not doom the firm in the long run. High rents benefit the politically connected and have been a big success in the eyes of the electorate (e.g. because of the high profits and share prices).

Finally, as Machado (1994) has succinctly documented, it is likely that a substantial share of the rents has been appropriated by Mitsubishi through a very self-serving TTA with its Malaysian partners. The excessive capture of Proton's rents by Mitsubishi and Japanese components suppliers may have limited Proton's capacity to re-invest productively. As noted earlier, poor governance and little emphasis on the development of local support institutions and productive capacities has allowed Mitsubishi and the Japanese suppliers to retain control of the key technologies involved (Rasiah, 1999). Hence, all four explanations of why the government has not been reducing tariffs appear plausible.

Within the framework to encourage training established by the Human Resources Development Council, Proton contributes 1 per cent of its payroll to its Fund, and has reclaimed close to 90 per cent of this for approved training schemes from the time of the application of the relevant act in 1993. Interviews, however, suggest that a significant share of the expenses incurred have been spent on non-technical motivational courses. Individuals still visit Mitsubishi Japan for training, but only to acquire assembly know-how for new models, e.g. a batch of trainees returned at the end of 1995 with new know-how for assembly of the Perdana model. In 1996, the firm opened a training centre on its premises, introducing modular training programmes. All ten workers interviewed claimed substantial learning acquired through participation in such programmes. During 1997–8, Proton workers reported serious problems with management following the downturn, which saw a big reduction in overall salaries as overtime work vanished and surplus workers were either required to do odd jobs or relocated to supplier firms. Many of these problems ended when the sharp rebound in the economy and the industry from 1999 raised demand for labour once again. Training has also expanded with the advent of the new Malaysian designed Proton Waja and the new engine being developed in Malaysia. The new Proton Waja models sold since 2000 have been powered by Mitsubushi engines for the 1.6 litre model and by Renault engines for the 1.8 litre model. According to Proton's annual report, new Malaysian-made engines are expected eventually to replace these imported engines, though no specific dates have been given.

Also, the utilisation of state-of-the-art process techniques has been limited. A private organisation was engaged in 1995 to introduce just-in-time practices in the production plant but problems led to the cancellation of the consultancy. Production managers reported continuing such efforts, but without any clear targets. The production cycle time in 1995 was 85 seconds, down from 125 seconds in 1985, suggesting that the pace of improvement has been rather slow, comparing quite unfavourably with the 60 seconds average cycle time achieved by German car assemblers (Roth, 1995). If the high quality standards of German firms are also taken into account, Proton's comparable cycle time may actually exceed 100 seconds.

Quality control circles (QCCs) have been introduced since 1985, but process improvements achieved through such small informal groups have been fairly modest. Minor improvements in pellet-unloading and jig-handling have been some of the improvements attributed to workers. The organisation of QCCs and their links with innovative learning (through access to state-of-the-art processes, and mechanical and auto documents) have not been very up to date. The relatively slow improvements in process technologies may be an important reason why delivery times to domestic consumers are still very much delayed – from one to three months.

Also, despite considerable co-operation from ethnic Chinese in EON's marketing and servicing networks, there has been a serious lack of co-ordination between such operations and manufacturing. The lack of effective co-ordination between the two has often left many sales outlets short of cars to meet domestic demand. Proton's production officials seem unable to explain the delivery delay of 1–3 months. In some cases, customers have only obtained the Waja model after 6 months. The rents associated with the Proton cars and the long waiting time have given rise to a lucrative black market. Also, six repair shop managers reported that their complaints of some frequent defects (e.g. involving automatic windows and central locking) have gone unheeded. A broad-based committee – involving representatives from all the firms and institutions – directly and indirectly involved in Proton's value-added chain could ensure more effective co-ordination to minimise market and government failures.

There is much more to be achieved – in terms of product and process technologies – for Proton to be classified a success. Given the challenges posed by recent liberalisation globally, a thorough review of Proton's performance is urgently required. The World Trade Organisation, formed in 1995, has set a period of five to eight years for lowering tariffs to 0–5 per cent, though it is unclear how this will be implemented. The ASEAN Free Trade Area (AFTA) has set a similar timeframe for the reduction of tariffs in Southeast Asia; AFTA's deadline for 'normal track' product trade liberalisation is also the year 2003 (ASEAN, 1995), though the Malaysian government has managed to extend the automobile imports deregulation dateline to 2005.

Whatever the impact of such trade liberalisation drives, Proton's operations – particularly its performance under circumstances of gradual reduction in protection – require more rigorous scrutiny so that the Malaysian economy eventually gains real benefits from the project. Unless such rigorous assessments and appropriate policy reforms are undertaken, the venture will continue to be a rentier operation benefiting only a few at the expense of Malaysian consumers and tax payers.

Conclusion

It can be seen that Proton has had three main objectives besides developing national automotive manufacturing capacity: first, to expand *Bumiputera* involvement in automotive manufacturing, second, to promote industrial linkages, and third, push the industry towards the technology frontier. The project has required the creation of rents to achieve financial viability for Proton and the generally inexperienced *Bumiputera* parts suppliers. While its launch initially involved little private–public sector co-ordination, largely due to poor inter-ethnic business co-operation in the early and mid-1980s, the persistence of this problem reflects poor governance. Ethnic Chinese sales and support networks have mushroomed all over the country as Proton car sales and services have become very lucrative. Such ethnic Chinese involvement, however, has not extended into manufacturing.

Proton has enjoyed substantial protection over fifteen years of existence, apparently even after achieving minimum efficiency (MSE) levels for manufacturing operations. Such factors as the need for R&D investments to catch up technologically and to shift more of the car value-added chain to Malaysia, the lack of effective government measures to review and improve performance, the influence of the politically connected, and the restrictive TTA conditions imposed by Mitsubishi (which has led to a significant share of the rents being transferred abroad) all appear to explain the persistence of high monopolistic rents. While investment in R&D may be considered desirable, the remaining three factors are unproductive. For Proton to become economically viable, the terms of the TTA should shift substantially to favour the local partners. The government should not only improve its bargaining capacity, but also expand local absorptive capacities. Hence, although production has expanded relatively strongly, a comparatively high share of the profits generated by Proton have continued to remain in the hands of its foreign technology suppliers.

The real test for Proton will come when protection is reduced, though greater liberalisation may pose a serious threat to its very existence. Even if liberalisation threats evaporate, it is important for the government to ensure that Proton is indeed moving towards the technology frontier and greater international competitiveness, and that domestic costs eventually decline below import prices. Otherwise, the Malaysian economy will continue to lose economic resources with little long-term gain, while domestic consumers continue to contribute to most of Proton's profits.

Palm oil refining: policy, growth, technical change and competitiveness

The emergence and dramatic growth of the palm oil refining industry has been a remarkable achievement in the industrial development of Malaysia. Exports of processed palm oil (PPO) products from the industry grew at an average compounded rate of about 25 per cent per annum over the past two decades. Currently, the industry, with an estimated annual refining capacity of about 11–12 million tonnes of feedstock, processes about 8–9 million tonnes of crude palm oil (CPO) and crude palm kernel oil (CPKO) yearly. This is an estimated 60 per cent share of world refined palm oil products and about 10 per cent of the major refined oils and fats.

The most important policy instrument used to promote the growth of this export-oriented industry was duty exemptions for the export of higher value-added processed palm oil products. Besides export duty exemptions, other tax incentives were also given to encourage the growth of the industry as part of a broader strategy of promoting resource-based industrialisation. Such provision of incentives were viewed by many in industrialised countries as subsidies to the industry, without which rapid growth in processing capacity and exports, its financial profitability and ability to compete in the world market could not have been sustained.

Todd (1978) found palm oil refining and fractionation in Malaysia socially and economically unprofitable during 1975–7. He argued that 'the rapid growth of the Malaysian processing industry and the somewhat disappointing returns on processed palm exports can be explained as effects of Malaysian Government subsidies'. Todd implied that the 'subsidies', in the form of export duty exemptions on PPO products and investment tax incentives, contributed to the high domestic financial profitability and attractiveness of investment in the industry. However, he argued that this was, on the whole, socially unprofitable given the export prices of palm oil products. He suggested that the rate of capacity expansion be slowed down and that more resources be put instead into marketing processed products.

Since then, there has been no other study examining the social and economic profitability and international competitiveness of the Malaysian palm oil refining industry. In contrast to the 1960s and 1970s, there have been several published case studies of firms/industries in the 'more advanced' developing countries that have undergone processes of technological learning and change in the 1980s and 1990s. These processes have led to higher productivity and improved levels of competitiveness in these firms/industries.

Considering the importance of the palm oil refining industry in the Malaysian manufacturing sector, these issues raise several pertinent questions. Was the industry subsidised and socially unprofitable, as claimed by Todd? While Todd's analysis was not static, it only covered a period of less than two years, which is too brief to adequately capture any technological and industrial learning processes, and resulting changes in technology and competitiveness that may have occurred in the Malaysian palm oil refining industry. This raises the question of whether the industry has made much progress in terms of competitiveness in the longer term.

Have there been significant technological learning and change? If so, have these processes resulted in the industry's capacity to compete in the world market, i.e. achieve international competitiveness? To answer some of these questions, this case study examines the competitiveness of the industry over a much longer time period than Todd, and attempts to identify various factors underlying the changes that have taken place.

This case study begins by examining the rapid expansion of the palm oil refining industry in Malaysia since the mid-1970s, and the policy environment in which this rapid expansion took place. It then examines, in the light of Todd's analysis, how competitive the refining industry has been during the 1980–94 period, using the concepts of 'gross refining margin' and 'competitiveness ratio'. Finally, it looks at the impact of policy incentives, growth, competition and other factors on the technical and structural changes in the industry and on its competitiveness.

Growth of the palm oil refining industry

The palm oil refining industry emerged as a significant industrial sector in Malaysia in the late 1970s. Prior to 1975, there were only a few factories refining and fractionating palm oil, palm kernel oil and coconut oil, mainly for the manufacture of cooking oil, margarine, vegetable ghee and soap products for the domestic market. In 1974, for instance, these factories were refining and fractionating about 90,000 tonnes of CPO, or less than 10 per cent of total CPO production. By 1976, 15 refineries – with an estimated capacity of 800,000 tonnes – were in operation, processing about 580,000 tonnes of crude palm oil, making Malaysia the country with the largest palm oil refining industry in the world. By 1977, the industry had a refining capacity of about one million tonnes and processed about 890,000 tonnes of crude palm oil. This also involved a new structural feature in the refining of oils and fats and their trade – the large-scale bulk refining of a single crude oil feedstock and large-scale bulk shipment of its refined products for export.

Refining capacity

Total approved capacity of operating refineries increased from 2.879 million tonnes of CPO feedstock in 1980 to a peak of 10.515 million tonnes in 1991, decreasing to 8.879 million tonnes in 1993, but increasing again to 10.013 million tonnes in 1994. This did not include the capacities of refineries that had ceased operation for one reason or another. Idle capacity of non-operating refineries has fluctuated but been persistent since the early 1980s. The total approved (operating and non-operating) capacity of such refineries increased from 2.879 million tonnes of CPO feedstock in 1980 to a peak of 13.007 million tonnes in 1987. Since then, it has fluctuated between 10 to 12 million tonnes of CPO feedstock. The expansion of fractionation capacity has also exhibited trends similar to those for refining.

The total number of (operating and non-operating) refineries increased from 45 in 1980 to peak at 57 in 1986, before declining to 46 in 1989 (as the licences of obsolete refineries were withdrawn), before increasing to 54 in 1992 (as new refineries were established). The number of operating refineries peaked at 51 in 1982, having risen from 45 in 1980, but declined drastically to 35 in 1984. Since then, operating refineries have numbered between 37 and 41. While operating capacities were increasing in the 1980s, the decreasing number of operating refineries is indicative of the increasing scale of refining operations in Malaysia and of the economies of scale in bulk refining CPO. In the 1990s, however, smaller refineries were successfully established to process crude palm oil and kernel oil into speciality fat products.

Exports of processed products

Consonant with the increase in refining and fractionation capacity, exports of processed (refined and/or fractionated) palm oil products (including palm fatty acid distillate) increased from 0.215 million tonnes in 1975 (when they were first exported in significant quantities) to 2.074 million tonnes in 1980. The volume of PPO exports increased further to 5.634 million tonnes in 1990 and to 6.595 million tonnes in 1994. Processed palm oil (PPO) exports grew at a compounded annual rate of 19.7 per cent over the twenty-year period. The share of PPO exports in total palm oil product exports increased rapidly from nil in 1974 to 18.4 per cent in 1975 and 91.3 per cent in 1980, 98.4 per cent in 1990 and 99.2 per cent in 1994. Total palm oil exports grew at compounded annual rates of 15.2 per cent, 18.9 per cent, 9.7 per cent and 3.8 per cent in the 1960s, 1970s, 1980s and 1990s (up to 1994) respectively. Exports of processed palm kernel oil also increased rapidly – from 38,971 tonnes in 1984 to 411,046 tonnes in 1994, increasing the export share from 10.0 per cent to 89.1 per cent of total palm kernel oil exports.

There has been a distinct change in the pattern of trade in palm oil products over the years. Traditionally, when exports from Malaysia were in the form of CPO, the major markets were the developed countries, particularly in Europe. With expansion of PPO product exports from Malaysia in the 1970s and 1980s, exports shifted to developing countries, particularly China, India, Pakistan and West Asia.

In summary, the domestic-oriented refining industry, with a capacity of less than 40,000 tonnes of crude oil feedstock in the early 1970s, grew into a large export-oriented industry with a capacity of about 12 million tonnes within a period of less than two decades. The industry currently processes about 8–9 million tonnes of crude palm oil (CPO) and crude palm kernel oil (CPKO) yearly, or 99 per cent of domestic production. This is an estimated 60 per cent share of world refined palm oil products and about 10 per cent of the major refined oils and fats.

What were the factors that contributed to this rapid expansion in palm oil refining capacity and exports in the 1970s and 1980s? Was this expansion achieved at the expense of economic welfare due to protectionist policies and subsidies, as claimed by Todd (1978)? Or was it a consequence of the inherent international competitiveness of the refining industry? In the next section, we examine the policy environment in which this rapid expansion took place for some of the answers.

Policy environment

In the 1960s, industrial policies and incentives were mainly geared towards import-substitution industries. Thus, during this period, the refining industry was mainly oriented to meeting domestic consumption. This outlook changed after the country adopted a more export-oriented industrialisation strategy from the late 1960s with various new policies and incentives introduced to promote such investments. Resource-based industrialisation was an important component of this strategy. Incentives were given for the establishment of industrial plants to further process domestically-produced raw materials – rubber, palm oil, timber and petroleum – and to increase domestic value-added in these export commodities (January 1982).

Prior to 1975, there were few attempts to undertake refining and fractionation of palm oil and palm kernel oil for the export market. The largely foreign- (mainly British-) controlled plantation companies preferred to maintain the exports of CPO produced on their oil palm estates. Similarly, multinationals (from the North) did not see much gains in (re-)locating their vegetable oil processing facilities in Malaysia. As late as 1978, the refining and fractionation of CPO in Malaysia for export was, on balance, seen as having limited potential (Dunn, 1978; Todd, 1978; Khera, 1978). Many reasons were advanced for the limited viability of local processing of CPO and the export of PPO products to major importing countries, such as in Europe:

- Malaysian refiners would be less efficient in processing CPO and manufacturing fat products because they had very little experience compared to refiners in industrialised countries in sourcing and processing crude oils and fats, and in blending, manufacturing and marketing fat products.
- Transportation, handling and shipping facilities and procedures in Malaysia were designed for the bulk movement of CPO for export. Modifications and additional facilities were needed to handle and transport processed palm oil products and to prevent quality deterioration as well as to meet standards. These would increase the cost of transportation of (processed) palm oil products from Malaysia to the importing countries.
- Processed products shipped from Malaysia to importing countries would be less acceptable because of quality deterioration due to transport and handling over long distances and periods. On arrival, the processed palm oil products would be of poorer quality and would require re-refining before being further processed into consumer products.
- High import duties on processed palm oil products in industrialised countries, especially in Western Europe, shaped the global oils and fats market, discouraging the import of processed palm oil products. These duties protected the local refining industry in importing countries in order to capture the higher margins derived from producing and marketing higher value-added consumer products for their domestic markets and for export to third countries.
- The supply and availability of processed palm oil products of specific qualities

as raw materials for a variety of blends and products would be adversely affected by the long distance and reduced interaction between suppliers and purchasers.

- The marketing of processed palm oil products in a highly substitutable oils and fats market was quite sophisticated. Only industrialised countries importing and processing CPO had the experience to market refined and fractionated palm oil products domestically and to third countries importing oil and fat products from these developed countries.

In retrospect, the arguments against the exports of PPO products from Malaysia proved to be exaggerated and one-sided and the Malaysian refining industry grew rapidly in the late 1970s and 1980s, much to the chagrin of refiners in Europe. Whatever the truth of these arguments at that time, they suggest that there were major barriers and obstacles to establishing an export-oriented palm oil refining industry in Malaysia. In such a scenario, policy incentives not only were necessary, but had to be sufficient, to attract the investment to develop such an industry.

The most important policy instrument used to promote the growth of an export-oriented palm oil refining industry in Malaysia was duty exemptions on exports of higher value-added processed palm oil products. Initially, the export duty structure was simple. From 1968, all PPO product exports were free of duty while a duty was imposed on CPO exports. In 1976, a more complex export duty structure was formulated to encourage more than the first stage of CPO processing. Thus, the duty structure on palm oil exports became a complex one, and there have been some changes over the years to rationalise it. In essence, high export duties were imposed on CPO and on aggregated categories of processed palm oil based on their respective prices.

However, processed palm oil products, which fell into five categories depending on the degree of processing, were allowed varying levels of export duty exemption. The amount of duty exemption increased (and hence, the export duty payable decreased) as the degree of processing and the value of the processed product increased. The export duty payable decreased to nil for the final fully refined and fractionated product category (MIDA, 1985). The intended effect of the export duty structure was to reduce the domestic prices of the crude and the less processed palm oil categories further away from their world prices while maintaining the prices of the more processed palm oil products nearer or at world prices. This encouraged a greater degree of CPO processing by increasing domestic processing margins (above world margins) down the processing chain. The net effect of this would be processing subsidies to domestic refiners transferred from the CPO producers.

Other incentives used to promote the industry have been tax relief and allowances for investment and export. The more important of these have included pioneer status, investment tax credits, export allowances, overseas promotion, public sector research as well as research and development incentives involving various kinds of tax exemptions. There have also been pre- and post-shipment export credit refinancing assistance programmes (MIDA, 1985).

By the mid-1980s, most of the tax incentives for basic refining and fractionation operations were withdrawn or had lapsed. The major incentives that remained were the export duty exemptions on refined and fractionated palm oil and palm kernel oil products. Tax incentives were increasingly only provided for activities further downstream, such as the manufacture of consumer and speciality fat products and oleo-chemicals (MIDA, 1985). Export credit refinancing assistance programmes were expanded in the 1990s, while tax deductions for overseas promotion and R&D activities continued.

Besides these direct incentives for investment in and export of PPO products, legislation was also introduced for the creation of institutions to assist the industry in R&D, training, and market promotion. In the late 1970s, the Palm Oil Research Institute (PORIM) and the Palm Oil Registration and Licensing Authority (PORLA) were established. PORIM has been responsible for research on all palm oil related activities, including the chemistry and technology of processing. PORIM has also been involved with PORLA and the Ministry of Primary Industries in technical and market promotion of processed palm oil products. In the mid-1980s, to counter the US soybean lobby against palm oil, the Malaysian Palm Oil Promotion Council (MPOPC) was also established to assist the industry in consumer-oriented promotional campaigns world-wide. All these organisations were supported by cesses imposed on the industry.

Besides incentives, there have also been important regulatory elements in the policy environment. The most important involved the monitoring and control of investments and capacities in the industry via conditions attached to the issue of manufacturing licences under the 1975 Industrial Co-ordination Act (ICA). The most important of these conditions were the (approved) maximum capacity and export limits, while less important conditions involved local content/material utilisation, employment, location and equity structure. Implicit in this regulatory framework is the monitoring and control of total installed refining and fractionation capacities to ensure that refiners have adequate CPO supply.

In 1986, however, approved maximum annual capacity conditions on refining and fractionation licences were relaxed, which led to a surge in capacity expansion among existing plants. Conditions for the issue of new licences for refining and fractionation plants were also relaxed. This relaxation on policy on the issue of new licences did not last, as refiners clamoured for greater controls as capacities greatly exceeded supply of CPO and capacity utilisation rates declined in the late 1980s – involving considerable socially wasteful, excessive competition. In the late 1980s, public policy making was strengthened and policies were better co-ordinated. In practice, the major incentives have been for investments and higher value-added export promotion, while incentives and institutional assistance for R&D, training and market promotion have been less significant and focused.

It is clear from the above that the Malaysian government provided a policy environment and substantial incentives conducive to the growth of the palm oil refining industry. These imply that there has been a welfare loss as a consequence for CPO producers and that the industry would not have been able to compete in the world markets without such intervention. In the next section, we examine the international competitiveness of the palm oil refining industry in the 1980s and 1990s by introducing the concepts of 'gross refining margin' and 'competitiveness ratios'.

Competitiveness analysis of the refining industry

To analyse the competitiveness of the Malaysian palm oil refining industry, the following gross margin estimations were made:

- monthly domestic gross refining and fractionation margin;
- monthly border gross refining and fractionation margin.

As there may be some doubt as to the extent to which prices at the Malaysian border reflect world (shadow) prices, we also substituted them with price estimates at the European border from a different source of data. We thus estimated gross margins for refining and fractionating palm oil at the European border in order to compute the competitiveness ratio of Malaysian refiners in relation to border prices in Europe. For a more complete analysis of competitiveness, we also decided to compare the gross margins for refining and fractionating palm oil within two major borders, i.e. Malaysia and the European Union. This was done by estimating the gross margins Malaysian refiners would have made if their refining and fractionation operations had been translocated to Europe. Owing to the limited data available, the period for the analysis was restricted to 1980–94. Data for the competitiveness ratio analysis using European border and domestic prices were limited to 1985–94.

Data sources and estimation

To analyse the competitiveness of the industry at country level, gross refining and fractionation margin computations need to use aggregate (average) price and aggregate product yield data for the country. For comparison, boundary conditions for determining the prices of products and raw materials have to be specified. Based on the availability of data, simplicity and comparability, we have defined prices to be on a 'delivered' or 'ex-' basis at refinery or port, depending on whether they represented domestic or border prices of CPO and PPO products.

Aggregate product yield data would vary from country to country, and from time to time, depending on the aggregate technology employed in the refining industry and the quality of feedstock used in the country at any particular time. Aggregate yields of PPO products can be estimated if data on crude, processed and refined palm oil products produced are collated and published on a regular basis. As such data were not available, case studies were used to estimate the yields of refined and fractionated products from CPO.

Prices

There are two major sources of price distortions in the international market for CPO and PPO products. The first is the export duty structure in major exporting countries, such as Malaysia, which have a decreasing export duty on PPO products as more value is added to CPO. The second is the import duty structure in major importing countries, such as the European Union (EU), which have a higher import duty on PPO products than on crude oil, i.e. through import tariffs which escalate with processing. These tariffs distort the prices of crude and processed products in exporting and importing countries, as well as world (border) prices.

The removal of tariffs would result in new equilibrium relationships, particularly in the spread between CPO and PPO products. However, these new price relationships are unlikely to make a significant difference to our competitiveness analysis. We assume that border prices for palm oil products, as expressed by the tariff-ridden equilibrium relationships, generally reflect shadow world prices (or at least their trends) under free trade conditions.

The data for actual transacted domestic (delivered) prices and border (free on board (fob)) prices for palm oil products in Malaysia were from Palm Oil Registration and Licensing Authority (PORLA) and Palm Oil Refiners' Association of Malaysia (PORAM). Where such data were not available, prices were estimated from (cost, insurance, freight ((cif)) prices in North-West Europe and appropriate freight, insurance and other handling costs and/or scheduled tariff rates. Border prices for palm oil products in Europe were the representative lowest asking (cif) prices in north-west Europe collated by *Oil World*. Where such data were not available, prices were estimated from Malaysian (fob) prices and appropriate freight, insurance and other handling costs. Domestic prices for palm oil products in the EU were estimated from cif (border) prices plus the scheduled import duties.

Product yields

Product yields from refining and fractionation depend on the technology employed for the purpose. The quality of the CPO feedstock is also an important variable determining the yield of refined and fractionated palm oil products. The refined and fractionated product yields of Malaysian refineries have improved over the period analysed. The major sources of improvements in refining yields have been optimisation in pre-bleaching, de-acidification and de-odourisation unit operations, and improved bleaching earth quality in the 1980s. With the introduction of high-pressure membrane filtration – replacing vacuum filtration – in dry fractionation, the product yield of the higher value olein fraction improved significantly.

The yield data for refined palm oil products were based on physical refining, the most common refining process since the late 1970s. The yield data for fractionated palm oil products were based on dry membrane fractionation, also the most common process since the mid-1980s. Product yields were estimated from regression analysis of actual production and quality data from several refineries in 1985 (Gopal, 1988). From the data, the following product yields (based on CPO feedstock) were used in gross refining and fractionation margin computations:

- RBD Palm Oil: 94.7 per cent
- RBD Palm Olein: 71.0 per cent
- RBD Palm Stearin: 23.7 per cent
- Palm Fatty Acid Distillate: 4.3 per cent.

Improvements in product yields, resulting from the technical changes in refining and fractionation over the years, were not considered in the computation of gross margins. Product yield improvements have a significant effect on absolute gross margins. However, the impact of product yield improvements can be qualitatively assessed and would not have affected trends in gross margins. More importantly, it would have had a similar effect across all gross margin estimations and, therefore, its impact on (comparative) competitiveness analysis is minimised.

The product yields used in the estimation of refining and fractionation margins at all the four 'locations' were the same as those estimated for Malaysian refiners. This was because the competitiveness analysis compares the gross margins Malaysian refiners would have obtained from their refining and fractionation plants (technology) and from CPO feedstock if they had been translocated to the border of another country, using the prices of the PPO products and CPO feedstock at these locations. However, the quality of the CPO feedstock and PPO products to and from these 'translocated' plants and product markets can be different, depending on the origin of the CPO feedstock and the markets for the PPO products. This would affect product yields and prices to some extent and, hence, gross margins.

Gross margins and competitiveness ratios

Value-added (i.e. the gross margins net of intermediate input costs) in refining and fractionation activities for both the domestic market and at the border would reflect trends similar to their respective gross margins, though more moderately. This is because material input costs for refining and fractionation have declined during the 1980s (with much lower material input requirements per unit of processed CPO in the late 1980s) as a consequence of optimisation and technical changes in palm oil processing. Using value-added, instead of gross margins, would not have significantly altered the results of the competitiveness analysis.

Examination of the annual average gross margins for refining and fractionating palm oil using Malaysian technology and CPO feedstock is very instructive. The negative and very low margins at certain times can be partly attributed to the price distortions on CPO as a consequence of the import and export duty regimes in trading palm oil products. Another contributing factor has been the rapid expansion in Malaysian refining and fractionation capacity, which led to the relatively high demand for – but uncertain supply of – CPO in the world market as well as the narrowing of the border price spread between PPO products and CPO.

Gross refining and fractionation margins in Malaysia declined tremendously during 1980 and 1994. Although there were some fluctuations, the annual average margins declined from about US$100/tonne CPO feedstock in 1980 to US$11 in 1991, but then increased slightly to about US$20 in 1993–4. In contrast, world margins for refining and fractionation were on an uptrend trend from about US$20/tonne for CPO feedstock at the Malaysian border in the early 1980s to about US$30 in the early 1990s. The competitiveness ratios for refining and fractionation of palm oil in Malaysia improved from more than 200 per cent of world gross margins in the early 1980s to less than 60 per cent in the early 1990s (based on Malaysian border prices).

Comparing palm oil refining and fractionation in Malaysia with world border prices in Europe clearly supports the earlier finding that the Malaysian palm oil refining industry was highly competitive, with a lower gross refining and fractionation margin than the world. From a comparison with the EU, which was protected by escalating import duties between crude and processed palm oil products, the ratios also suggest (although again, no clear trend is discernible) that palm oil refining and fractionation activities in Malaysia were highly competitive, with a gross margin of less than 30 per cent using European domestic prices. This is a vivid indication of the efficiency, technological progress and high level of international competitiveness achieved by the Malaysian palm oil refining industry. Through processes of industrial growth, competition, entrepreneurship, technological learning as well as ensuing technical and structural changes over the years, the industry achieved a high level of international competitiveness in the late 1980s, which is the subject of examination in the next section.

Competitiveness: the key factors

This section attempts to examine the factors that contributed to the high level of competitiveness achieved by palm oil refining and fractionation activities in Malaysia.

Policy incentives and growth

As shown in the section on the policy environment, the Malaysian government employed an array of policy instruments to promote an export-oriented palm oil refining industry. The primary objective of the policy incentives was to attract investments for processing CPO and exporting PPO products, by creating an environment in which the (private) financial profitability of such investments would be high. This was clearly achieved very successfully, as described in the section on growth of the industry. Malaysian refining capacity grew rapidly at a compounded rate of about 25 per cent per annum over the two decades since 1974, processing in 1994 more than 8 million tonnes or 99 per cent of the CPO and CPKO produced.

The policy incentive mainly responsible for this rapid expansion was the duty exemptions on PPO exports. The high export duty on CPO ensured that the domestic prices of CPO were well below world prices. With lower or no export duties as a result of export duty exemptions, PPO products could be exported at or near world prices. With such relative prices for PPO products and CPO, refiners received higher domestic margins, and were guaranteed bigger profits for processing CPO and exporting PPO products. At this time, the main requirement for realising huge profits from processing CPO was finding lucrative markets for PPO products. Tax relief and allowances for profits and exports have further increased the private profitability of these processing activities.

The creation of a highly profitable environment by providing generous incentives was crucial to the establishment and growth of the palm oil refining industry in Malaysia. In the absence of these incentives, the industry would not have been established because there were several major obstacles to its establishment. As argued by Todd (1978), Malaysia lacked (static) comparative advantage in palm oil refining and fractionation activities. Similarly, the industry lacked comparative advantage in marketing and distributing PPO products overseas (due to the lack of infrastructure, facilities and capabilities, resulting in high costs to undertake them). Further, there have been trade (escalating tariff and non-tariff barriers), commercial (restrictive marketing and business practices) and technical (product, transportation and quality problems) barriers to imports of PPO products.[8] In the face of these obstacles, generous policy incentives have been necessary to ensure high financial profitability and reduced risks for investment. Even then, significant investment flows into CPO processing for export followed six years after the export incentives were introduced in 1968. Needless to say, the response of foreign direct investment has been far more muted throughout.

As the pioneer investors began realising huge profits at relatively low risk in CPO refining and fractionation activities by the mid-1970s, investments in the industry snowballed. Refining capacity and processed palm oil exports in the early period (1974–9) grew rapidly at more than 60 per cent per annum when the generous incentives maintained a highly profitable environment and CPO production (supply) continued to be in excess of refining capacity. In this respect, the increases in CPO supply was an important factor in maintaining this rapid growth, as CPO production doubled during this period. Ironically, however, the generous incentives provided for investment were also the seeds for the erosion of high financial profitability. However, erosion of the high domestic margins and profitability was, in turn, a blessing in disguise, as it spawned a competitive environment that pushed the industry towards greater efficiency and productivity.

Growth, competition and gross margins

From our analysis in the earlier section, growth in the competitiveness of the refining industry was a consequence of the ability to reduce gross domestic margins. Two key processes can be said to have contributed to this reduction in gross domestic margins in two more or less overlapping phases. The first phase (1977–83) can be characterised as the period when excess profits in refining and fractionation activities were eliminated. As investments rolled in, lured by the huge profits of the early and mid-1970s, increases in refining capacity greatly exceeded increases in CPO supply. This led to increasingly higher domestic demand for CPO and relatively higher CPO prices, nearer to world border prices, thus reducing gross domestic margins, eliminating excess profits, and narrowing the large differences with border margins.

Domestic demand for CPO was further exacerbated as total domestic refining capacity outstripped total domestic CPO supply at the turn of the decade. And as capacity continued to expand in this second phase (1980–8), it created intense competition for CPO supplies, pushing CPO prices closer to and then above world border prices by the mid-1980s. This resulted in the reduction of gross domestic margins below that of world margins. The most important distinguishing feature of this phase – in comparison with the earlier phase – was that technical, organisational and structural changes enabled the industry to cope with the reduction in gross domestic margins.

Evidence on gross domestic margins in comparison with the export duty on CPO and total approved capacity of installed refining plants indicates that the export duty on CPO contributed to lower domestic prices for CPO than the world border price, and to higher domestic margins until around 1983. Based on theoretical calculations, the domestic margin should have been higher (in most instances) than the export duty payable on CPO (as the refined-fractionated products are duty-free) by an amount equivalent to the border margin. In practice, however, it was lower than the export duty on CPO. Hence, the effectiveness of the CPO export duty (in reducing domestic CPO prices by an equal amount below border prices) was already declining in the early 1980s (from which time data has been available). With that, domestic margins were generally also in decline compared to the early years of rapid growth of the refining industry in the mid-1970s.

Since 1986, the export duty has had no impact at all on gross domestic margins. Technological, organisational and structural changes thus became critical for the survival of firms and a significant number of inefficient firms had to cease operations. As capacity increased dramatically in 1986 and 1987, the industry experienced lower capacity utilisation and intense competition for CPO supplies. The CPO export duty became totally ineffective in lowering domestic prices for CPO (below the 'world' price), and in thus maintaining relatively higher domestic margins than at the border. Instead, the high domestic demand for CPO pushed domestic prices above world border prices, reducing domestic margins below border margins. These reductions in domestic margins, which were crucial for the industry achieving a high level of international competitiveness, were made feasible by the greater efficiency and productivity generated by investment and capacity expansion in the mid-1980s. By then, the industry had sufficiently accumulated the necessary entrepreneurial, marketing and technical skills to bring about major technical changes, improve efficiency and sustain profitability. The next section examines, in more detail, this second aspect of the reduction of gross domestic margins in Malaysia.

Technical and structural changes

The second phase of reduction of domestic gross margins was the result of the technical and organisational changes in the refining, fractionation and export of palm oil products and the consequent changes in the industry's structure. The technical changes included: the modification and optimisation of refining and fractionation unit operations – the switch from chemical to physical refining, higher plant throughput (capacity stretching) and reduction in steam consumption by de-acidifier/deodoriser modification, heat recovery and segregation of crude oil by quality, reduction in bleaching earth and phosphoric acid dosage, with better knowledge of oil and earth quality and their interactions, higher fractionation yields at lower cost with the switch from solvent fractionation and dry vacuum filtration to high pressure membrane filtration; better control over the variability in product quality with better knowledge of the impacts of CPO quality, processing and transportation conditions on final delivered quality; economies of scale with bulk refining, fractionation and export; greater quality control and efficiencies in the bulk transport and handling of products for shipment; and localisation of equipment design and manufacture.

These changes were driven by the rapid growth in capacity and the large size of the industry. The huge expansion in capacity in the late 1970s and early 1980s resulted in intense competition for CPO, as capacity exceeded domestic CPO supply. Refining margins and profits were squeezed as a consequence of higher bids for CPO. With the erosion in (excess) profits in the 1980s, refiners had to improve efficiency to earn reasonable profits. With the successive installation and operation of an increasing number of refining and fractionation plants, the industry accumulated a lot of experience, skills and knowledge in refining and fractionation technology and the processing of CPO.

At the same time, the large size of the industry, with more than US$500 million of investment, created the necessary economies and incentives for strong backward linkages to domestic engineering and technology related activities (see the later section on technology imports and local capacity). Similarly, equipment vendors (mainly of foreign origin) also had an incentive to optimise refining and fractionation technology, considering the huge market potential as palm oil was the fastest growing edible oil in the world market. Local refiners collaborated with vendors to improve and optimise CPO refining and fractionation technology, also involving greater localisation of equipment design, fabrication and installation.

Refining and fractionation technology was modified and optimised for continuously processing a single oil feedstock, namely CPO. The major sources for these technological improvements were the differences in the refining properties of CPO from those soft oils for which the technology was originally designed. The changes were made possible by the accumulation of knowledge and skills on: the physical and chemical properties of crude and processed palm oil products; the effects of crude palm oil quality and processing parameters on final product quality; and product specifications, product uses and market demand. This process of learning was assisted by public sector institutions (such as PORIM, the palm oil research institute) and private sector organisations (such as the refiners' association) which conducted technical and marketing research, promotion and extension activities.

With the liberalisation of regulatory controls on capacity expansion in 1986, refineries competed to establish large-scale plants with the newly optimised technologies and significant economies of scale. The expansion in capacity also led to the closure of smaller less efficient and unprofitable refineries under the low domestic refining and fractionation margin regime. This has been a persistent phenomenon since 1983, when capacity greatly exceeded CPO availability. In the 1980s and 1990s, there have also been take-overs of inefficient and non-operating refineries by plantation groups, investors with potential market niches and other more efficient refineries, which have led to capacity upgrading and expansion.

These technical and structural changes have created a new and unique industrial structure in the refining and fractionation of CPO in Malaysia – involving highly optimised, continuous large-scale bulk refining and fractionation of CPO, and the commoditisation, bulk marketing and distribution (export) of PPO products – which has resulted in scale economies to competitively sell refined and fractionated products world-wide.

Foreign direct investment and local entrepreneurs

Foreign direct investment (FDI) played a very minor role in terms of equity in the development of the palm oil refining industry. By 1987, when investments in refineries had already peaked, foreign equity accounted for about 17 per cent of total paid-up capital (MIER, 1990: 9). In contrast to the general flow of FDI from developed countries into developing countries, most FDI in the Malaysian palm oil refining industry was not from the 'North', with most coming instead from India, Singapore and Hong Kong. The Indians have had a leading and key role in the Malaysian palm oil refining industry, with interests in as many as eight installed refineries in the 1980s. The main reason has been the considerable imports of processed palm oil products to India from Malaysia from the late 1970s until the late 1980s. Another factor could be foreign exchange controls in India then which encouraged Indian capital overseas to continue investing their profits abroad rather than be repatriated home.

From the more developed countries, only the Japanese invested significantly in refining and fractionating palm oil for export in the 1970s and 1980s, when policy incentives encouraged rapid growth. Whereas, despite the generous incentives, which resulted in high domestic financial profitability, Western companies showed no interest in investments in palm oil refining and fractionation for export. Their arguments against such investments have been described elsewhere in the section on policy environment. Unilever was the only Western multinational company that had interests in processing CPO in Malaysia in the 1970s. However, its refinery had been established in the 1960s under the import-substitution incentive programme to produce consumer oil and fat products for the domestic market. It was only two decades after policy incentives were first introduced that a large

American multinational oils and fats company acquired two operating refineries. This step was taken after recognising the need to have a stake in the highly efficient Malaysian palm oil processing industry to strengthen and boost their international oils and fats trading activities.

Plantation groups producing large quantities of CPO in Malaysia were expected to lead investment in the Malaysian palm oil refining industry. However, because the dominance of British interests in the major plantation groups and their negative view of the local potential for palm oil processing, the responses from these groups fell far short of expectations. The plantation firms that pioneered establishing local refining capacity in the 1970s were mainly controlled by local interests, from both the private sector as well as public sector agencies. Others who invested significantly included the local refiners who had been processing CPO on a small-scale for domestic consumption and independent investors who had knowledge of the industry and of the profits that the export incentives could generate or were generating for refineries already in operation.

High profits, including rents from local CPO processing as a result of the generous incentives, seem to have attracted excessive investments, with unfavourable short-term consequences but positive medium-term implications. Entrepreneurship in the industry grew rapidly with intense competition, lower margins, higher risks and rapid accumulation of processing and marketing experience in the 1980s. This involved the competitive search for processing efficiencies, product differentiation and new markets. This process of rapid change was hardly smooth, with the more enterprising refiners becoming technology and market leaders, and the less successful often suspending operations in the low season when CPO supplies were tight. But with diffusion of new, non-propriety technology, there was overall improvement in the efficiency and competitiveness of the industry as a whole.

Technology imports and local capacity

Limited FDI and local experience encouraged investors in the 1970s to seek out refining and fractionation technologies from foreign sources. Several foreign equipment designer-manufacturers also competed to market their wares to these investors. However, early technology selection and adoption was less discriminating because of the lack of experience. The mainly foreign consultants and equipment vendors, selected by the investors, based their design of refining and fractionation plants on existing plants in the advanced countries, while adding the latest improvements.

Technology imports in the 1970s were high, with all major plant equipment being imported. Local capital expenditures were mainly for civil and structural works, fabrication of simple vessels, tanks and piping and the erection of the plant. However, technology imports gradually declined, and by the late 1980s, foreign capital expenditures were limited to sophisticated and precision plant equipment such as separators, high pressure presses, chillers, filters, membranes, motors, engines and control devices.

How was the localisation of technology achieved? For a developing country in the 1970s, Malaysia had a relatively high level of local skills in basic engineering. For instance, in the early 1970s, engineering was a major industrial sector, accounting for over 13 per cent of manufacturing value-added and growing at about 15 per cent per annum. Historically, agriculture and agro-based industries have had strong backward linkages to the local engineering industry. Engineering output to the primary commodities export sector (tin, rubber, oil palm and timber) was large, but declining in the sixties and 1970s, suggesting that engineering growth was increasingly driven by import substitution and then export-oriented manufacturing. The low level of vertical integration in refining and fractionation technology made local fabrication of less sophisticated equipment and components feasible. Such local fabrication was slowly upgraded to involve more complex equipment and components as more knowledge and skills were acquired through the processes of learning by doing.

Technology search, import, design, fabrication, project execution, as well as erection, commissioning, start-up, operation and maintenance of the pioneering refining plants provided training in a range of skills – from plant design, fabrication of plant hardware and equipment, plant erection to equipment installation involving production and maintenance engineers, supervision and operators. These processes of learning also contributed to the modification and optimisation of palm oil refining and fractionation technology, with current technology quite different from that of the first refining plants installed in the 1970s. With the successive installation and operation of a greater number of refineries, local skills have reached a level where local consultants, contractors, operating staff and equipment have been exported to other Asian and African countries establishing local oils and fats processing facilities since the 1980s.

Conclusions

The emergence and rapid growth of an export-oriented palm oil refining industry in Malaysia has been a remarkable achievement in its industrial development. Processing capacities grew at about 70 per cent in the 1970s and 15 per cent in the 1980s. In the 1990s, the industry processed 99 per cent of domestic CPO supply and produced 60 per cent share of the world's refined palm oil products. As the industry initially lacked (static) comparative advantage and faced various barriers to its exports, generous policy incentives were crucial for the rapid expansion in palm oil refining and fractionation capacities in the 1970s and 1980s. The most important policy incentives were duty exemptions on exports of PPO products (while a high duty was imposed on CPO exports) and tax relief as well as allowances on profits and exports. However, the Malaysian refining industry did not languish in the high-profit environment created by the generous incentives, as earlier studies have suggested. Instead, there were significant technical, organisational and structural changes in the 1980s, which created a highly competitive industry in Malaysia within a decade of its emergence.

To demonstrate the dynamism of the industry, changes in the level of international competitiveness during 1980–94 were measured using competitiveness ratios. Using the concept of effective exchange rate, the competitiveness ratio compared gross margins (or value-added) for processing CPO at domestic market prices with world border prices. The analysis indicates that the Malaysian refining industry improved dramatically from being internationally uncompetitive in the early 1980s to become highly competitive since the late 1980s. The gross margin for refining and fractionating palm oil in Malaysia decreased from more than 200 per cent of the world gross margin (based on Malaysian border prices) in the early 1980s to less than 60 per cent in the 1990s. Malaysian refiners in the 1990s competed internationally with costs of about only 30 per cent of the gross margin using European domestic prices.

The gains in international competitiveness were driven by the rapid growth in processing capacities, large size of the industry and domestic (inter-firm) competition. The rapid growth of the industry in the highly private profitable environment led to several developments. Ironically, one was the elimination of excess profits created by the policy incentives, as processing capacities increased faster than domestic CPO supply, leading to relatively higher CPO demand and prices. Competition for CPO supplies and price pressures intensified in the 1980s as domestic processing capacities outstripped CPO supplies.

This led to the search for greater efficiency and productivity. Intense competition made the realisation of a very high level of processing efficiency feasible. The growth and large size of the industry (as a consequence of policy incentives and increasing CPO supplies) generated significant economies and strong backward linkages to domestic engineering and technology related activities and also led to the accumulation of experience, knowledge and skills in CPO processing techniques and technology. Refiners collaborated with equipment vendors and manufacturers to improve and optimise processing technology – building on differences in the physical properties of CPO compared to the soft oils on which the initial imported processing technologies were based – and to localise the fabrication of processing equipment to a greater extent.

Eventually, domestic competition, technical changes and economies generated by the accumulation of experience, knowledge and skills and the growth of a large industry led to the development of a unique industrial structure – the highly-optimised, continuous large-scale bulk processing of CPO and the commoditisation, bulk marketing and export of PPO products. These factors gave rise to an industry with a competitive advantage in the world market for PPO products.

While policy interventions (incentives) were critical in overcoming political and economic obstacles to the establishment of an export-oriented refining industry, its direct impact on increasing the level of competitiveness was marginal. The most important policy intervention – the export duty structure – was effectively a 'market-based' transfer pricing mechanism involving CPO, intended to favour CPO refiners over CPO producers. During the infant stage of the CPO refining industry (when processing capacities were less than domestic CPO supply) in the 1970s, this mechanism had the intended effect of providing favourable (lower) CPO transfer prices and higher profits for refiners. The unfavourable (lower) prices for CPO producers, however, did not affect the expansion in domestic CPO production and supply as it was highly profitable in the 1970s and could easily absorb the lower CPO prices.

But since the mid-1980s, with CPO profitability declining and expansion in CPO production slowing down, more intense competition for CPO supplies and improvements in CPO processing efficiency in refineries reduced the domestic–world price margin, increasing domestic CPO prices and providing favourable (higher) transfer prices to CPO producers. These changes in CPO transfer prices enhanced the incomes and profits of both CPO refiners as well as producers (suppliers) when they needed it most and benefited the palm oil industry (producers and processors) as a whole. While the export duty interventions succeeded in expanding processing capacities and PPO exports, they did not contribute 'actively' to achieving higher processing efficiencies and gaining international competitiveness. However, the growth of the refining industry, as a consequence of the policy incentives, contributed to the more 'passive' processes of 'learning by doing'.

Other incentives meant to promote greater efficiency were found wanting. Double taxation relief for R&D and market promotion by themselves would not have been sufficient incentives to undertake activities in technology search, training, R&D in processing and products, and investments in improved technology. Similarly, institutional support has only provided basic technical and market information and forums in which the search for processing efficiencies could be based. Most of the gains in competitiveness were the result of competitive pressures to survive as well as increase or sustain profits and were complemented by entrepreneurial and technological skills accumulated by individuals in progressive firms actively pursuing growth and greater efficiency. This was particularly so for firms which were not backward integrated and had to purchase CPO supplies in the open market.

On hindsight, it is fortunate that regulatory controls over entry and capacity expansion were rather 'flexible' and largely deregulated by the late 1980s. Strict regulation of entry and capacity expansion to levels lower than domestic CPO supply would have maintained relatively lower CPO demand and prices and higher profits for CPO processing. This would have limited competitive pressures to improve efficiency and competitiveness. Further, the rationalisation of the industry through the acquisition of inefficient firms or those that had ceased operating, by other refiners, plantation companies or investors with market niches in the mid-1980s and after, were largely left to market forces.

The major costs of policy intervention, to the palm oil industry as a whole, were due to the less discriminating large-scale technology imports of the early years of the refining industry. These costs we believe have been more than offset by the gains in international competitiveness and other positive externalities generated by technological accumulation and greater indigenous technological capacity. However, better planning and policy interventions, such as those promoting more active and effective processes for technological learning as well as development of indigenous technological capacity, would have reduced technology imports and learning costs, and enhanced the international competitiveness of the industry more rapidly. The lack of FDI and involvement of multinational corporations may well have limited more rapid acquisition of technology, marketing skills and international levels of competitiveness. However, there has probably also been much more active search for technology imports and more effective processes of technology transfer and learning as a consequence.

Concluding remarks

Growth in Malaysia's manufacturing sector has been guided by two major strategies, namely import substitution (IS) and export-orientation (EO). The former has been constrained by the small domestic market, while the latter has benefited from access to global markets. Exports expanded easily as the foreign transnationals already enjoyed considerable global market shares prior to their relocation in Malaysia. The electric/electronic and textiles/garment sub-sectors – the two most export-oriented industries in Malaysia – are also dominated by foreign ownership.

IS and EO industries have both enjoyed distorted relative prices due to protection and subsidies respectively. IS rents attracted investments, especially from abroad, until the near saturation of the domestic market by the mid-1960s. Later, government-sponsored heavy industries expanded quickly from the late 1980s. Although it is difficult to establish the efficiency of government-owned heavy industries (as large debts incurred in their establishment have been written off), they did expand considerably to dominate the domestic market. EO rents have attracted foreign transnationals largely engaged in the processing and assembly of imported inputs for export. The sheer size of global markets has facilitated rapid expansion of EO industries, which have become leading generators of manufacturing employment, output and exports. A combination of rents – offered in the form of financial (especially tax) incentives, low wages, good infrastructure, political stability and government support – has attracted risky lumpy investments in export processing, and even in some design activities since the 1980s.

Like import substitution, export-orientation has also involved distorting relative prices. Initially driven primarily by employment generation considerations, IS and EO involved relatively less pro-active participation by the government until the 1980s. Given the government's relatively weak bargaining power *vis-à-vis* foreign capital and limited strategic economic vision from the late 1950s until the early 1980s, the government hardly supported the expansion of domestically owned productive capacity. Apart from schooling and infrastructure, the government invested little in the development of industry skills, research and development and other support for local manufacturing firms until the late 1980s.

From the mid-1980s, especially with the introduction of the IMP, the government assumed a progressively more interventionist stance in some regards, while withdrawing in other respects in line with its stated commitment to economic liberalisation, giving the overall impression of incoherent industrial policy. On the positive side, rents were increasingly tied to the development of domestic production capacity, rather than simply to investment and employment generation, as was the situation before the mid-1980s. Human capital, research and development, linkages, exports and technologically strategic manufactures have all become privileged, e.g. by enjoying tax incentives. The period since the mid-1980s has also involved new distortions by the government to use its improved bargaining position *vis-à-vis* foreign capital to encourage technological deepening.

Despite the government's pro-active role, existing industrial strategy in Malaysia also has some flaws. In at least three areas, changes will be necessary to further strengthen the manufacturing sector. First, weak links between IS and EO industries need serious attention. Enforcement of the 30 per cent domestic sourcing condition does not sufficiently address this problem. Large numbers of foreign suppliers have flocked to Malaysia, especially since 1986, securing supply contracts which local firms may otherwise have obtained (see Rasiah, 1993a). While foreign parts suppliers who use technologies well beyond existing local technological capacities (e.g. fabricated wafers and lead frames) are both desirable and unavoidable, studies show that several foreign firms are actually penetrating markets previously supplied by local firms (see Rasiah, 1990; 1993a). This issue may become more serious with the implementation of the common effective preferential tariff (CEPT) scheme from 1993, which aims to reduce trade barriers among the ASEAN economies. To prevent further undermining of promising enterprises in ASEAN, efforts should be taken to shelter them against unfair competition, especially in the initial phase.

Second, there needs to be a rigorous assessment of incentives offered to IS and EO firms to eliminate unnecessary carrots offered to firms so that only efficiency-enhancing, linkage-spawning and technologically-strategic firms are supported. Even for these firms, rents should be gradually withdrawn to force them to become internationally competitive. To achieve productive efficiency, rents that do not generate gradual improvements in competitiveness should be terminated. In this regard, there needs to be greater dissemination of performance data in the state-controlled heavy industries so that assessments can be made on their performance. While the stick should be applied to non-performing firms, the choice of products and industries for promotion should not only be limited to profitable ones as complementary effects may make non-profitable industries and products essential for the growth of profitable ones. Hence, the issue of complementary influences in industrial structural change is crucial.

Finally, rents should be tied to aspects of technological deepening such as greater emphasis on process and product research and development. Indeed, the manufacturing sector has reached a level where further expansion strongly depends on moving on to higher technological niches greatly dependent on greater investments in research and development. Only with more emphasis on innovation can manufacturing firms move on from being mere assemblers and sub-contractors to becoming original equipment manufacturers (OEM). Although initiatives are being taken to increase innovative research in Malaysia, there is no coherent strategy to implement them. The development of institutional networks to support technology development and dissemination in industries should seek to minimise costs and maximise gains.

While markets generally influence the establishment of large firms and institutions, policy analysis should also recognise and emphasise the potential of co-operation as an additional co-ordination mechanism. To strengthen participation and to achieve long-term efficiency gains, collaborative networks between firms and these institutions should be encouraged. While governance appears to have been important for Malaysia's industrial growth, this should not be interpreted to imply a total rejection of the market. The market has performed the useful role of generating resources and signalling current pecuniary influences. Moreover, like the market, government intervention also involves costs. Hence, a combination of visible and invisible governance should guide future industrial policy in Malaysia.

Notes

1 Krueger (1974) and Bhagwati (1978) have written on the efficiency dissipating effects of rent-seeking. They assume that market-determined allocation of resources will eliminate rent-seeking and thus generate optimal growth. (The insight that discount rates have to exceed interest rates to stimulate investment in risky innovative activities has become significant in new growth models (Romer, 1986; Lucas, 1988; Helpman and Krugman, 1989; Grossman and Helpman, 1991. Also, public goods command different demand structures and generate externalities with wider societal implications.) Also, due to the endemic nature of information asymmetries, quasi rents are bound to occur even in day-to-day transactions not related to scale economies (Coase, 1937; Williamson, 1979, 1985).

2 The Industrial Coordination Act of 1975 is unlikely to have been a major cause of the relative fall in foreign manufacturing especially export-oriented investment. MIDA (1988) promotional brochures stipulated *Bumiputera* equity participation only for firms (with paid-up capital beyond a certain level) selling more than 20 per cent in the local market.

3 See Krugman (1980; 1989) for an analytical account using neo-classical tools to demonstrate gains from IS as a strategy for promoting exports.

4 Unpublished data from C.B. Edwards.

5 See Machado (1994) for a lucid political economy account of Proton's formation and its localisation programme.

6 Machado (1994) has argued that the deal has seriously restricted Proton's ability to raise local productive capacity.

7 See also Machado (1994).

8 See also the sections on the policy environment and foreign direct investment.

References

Amsden, A. (1989) *Asia's Next Giant: South Korea and Late Industrialization*, New York: Oxford University Press.

Amsden, A. and L. Kim (1985) 'The role of technological capability in South Korea', mimeo, Development Research Department, Productivity Division, World Bank, Washington, DC.

Anuwar Ali (1992) *Malaysia's Industrialization: The Quest for Technology*, New York: Oxford University Press.

Anuwar Ali and Wong P.K. (1993) 'Direct foreign investment in the manufacturing sector', in Jomo K.S. (ed.) *Industrialising Malaysia: Policy, Performance and Prospects*, London: Routledge.

Aoki, M. (1986) 'Horizontal vs. vertical information structure of the firm', *American Economic Review*, 76(5): 971–83.

Arshad, Fatimah M. and R.A. Ghaffar (1989) 'Malaysia's primary commodities: constant market share analysis', in Arshad, F.M., *et al.* (eds) *Malaysian Agricultural Policy: Issues and Directions*, Serdang: Centre for Agricultural Policy Studies, Universiti Pertanian Malaysia.

ASEAN (1995) *AFTA Reader*, Jakarta: ASEAN Secretariat.

Balassa, B. (1979) 'The changing pattern of comparative advantage in manufactured goods', *Review of Economics and Statistics*, 61: 259–66.

—— (1982) *Development Strategies in Semi-Industrial Economies*, Baltimore, MD: Johns Hopkins University Press.

Bank Bumiputra Berhad Economic Review (1990) 'The Malaysian telecommunication industry', 9(4), July–August: 35–51.

—— (1995) 'Telecommunication industry: exciting times ahead', Jan.–March: 49–77.

Basiron,Yusof (1988) 'The export performance and competitiveness of Malaysian rubber and palm oil', National Seminar on Trade and Development, 29–30 March, Universiti Pertanian Malaysia, Serdang.

Bell, M., B. Ross-Larson and L.E. Westphal (1984) 'Assessing the performance of infant industries', *Journal of Development Economics*, 16(1–2): 101–28.

Bhagwati, J. (1978) *Foreign Trade Regimes and Economic Development*, Cambridge, MA: Ballinger.

—— (1985) 'Foreign trade regimes', in J. Bhagwati (ed.) *Dependence and Inter-dependence*, Oxford: Basil Blackwell.

—— (1988) *Protectionism*, Cambridge, MA: MIT Press.

Broadway, R., D. Chua and F. Flatters (1991) 'Tax induced investment distortions in Malaysia', Malaysian Institute of Economic Research, Kuala Lumpur, processed.

Chan, K.H. (1994) 'Program Lokalisasi Industri Kereta: Satu Kajian Pembuatan Kereta Proton Malaysia', undergraduate dissertation, Universiti Kebangsaan Malaysia, Bangi.

Chandler, A.D. (1962) *Strategy and Structure*, Cambridge, MA: MIT Press.

Chang, Ha-Joon (1994) *The Political Economy of Industrial Policy*, New York: St Martin's Press.

Chee, P.L. (1985) 'The Proton saga – no reverse gear! The economic burden of Malaysia's car project', in Jomo K.S. (ed.) *The Sun Also Sets*, 2nd edn, Kuala Lumpur: INSAN.

Coase, R.H. (1937) 'The nature of the firm', *Economica*, 4(16), November: 386–405.

Coffin, G., B. Larue, M. Banik and R. Westgren (1993) 'Competitiveness of the Canadian food-industry', *Canadian Journal of Agricultural Economics*, 41(4): 459–73.

Dahlman, C.J, B. Ross-Larson and L.E. Westphal (1987) 'Managing technological development: lessons from newly industrializing countries', *World Development*, 15(6): 759–75.

Daito, E. (1986) 'Recruitment and training of middle managers in Japan, 1900–1930', in K. Kobayashi and H. Morikawa (eds) *Development of Managerial Enterprise*, Tokyo: University of Tokyo Press.

Doner, R. (1991) 'Approaches to the politics of economic growth in Southeast Asia', *Journal of Asian Studies*, 50(4): 818–49.

—— (1992) 'Limits of state strength: toward an institutionalist view of economic development', *World Politics*, 44: 398–431.

Dunn, C.P. (1978) 'Economics of further processing: markets', Market Development of Palm Oil Products, Proceedings of a seminar organised by the International Trade Centre (ITC) UNCTAD/GATT of the United Nations, 27 March–1 April, Geneva: ITC, UNCTAD/GATT.

Edwards, C. (1992) 'Malaysian industrial policy in the 1990s – look East?' in R. Adhikari *et al.* (ed.) *Industrial Trade Policy Reform in Developing Countries*, Manchester: Manchester University Press.

Edwards, C.B. (1975) 'Protection, profits and policy: industrialisation in Malaysia', PhD thesis, University of East Anglia, Norwich, UK.

Ekelund, R.B., Jr and R. Tollison (1981) *Mercantilism as a Rent-seeking Society: Economic Regulation in Historical Perspective*, College Station: Texas A&M University Press.

Evans, P. (1987) 'Class, state and dependence in East Asia: lessons for Latin Americanists', in F. Deyo (ed.) *The Political Economy of the New Asian Industrialism*, Ithaca: Cornell University Press.

Fong, C.O. (1986) *Technological Leap: Malaysian Industry in Transition*, Singapore: Oxford University Press.

Fransman, M. (1982) 'Learning and the capital goods sector under free trade: the case of Hong Kong', *World Development*, 10(11): 991–1014.

—— (1986) 'International competitiveness, technical change and the state: the machine tool industry in Taiwan and Japan', *World Development*, 14(12): 1375–96.

Fransman, M. and K. King (eds) (1984) *Technological Capability in the Third World*, London: Macmillan.

Gopal, J. (1988) 'Refining margin variances and crude palm oil quality', *Proceedings of the 1987 International Oil Palm/Palm Oil Conferences: Progress and Prospects, Conference II: Technology, 29 June–1 July 1987*, Kuala Lumpur: PORIM.

—— (1996) 'Competitiveness of the palm oil refining industry in Malaysia', *Proceedings of the 1996 PORIM International Palm Oil Congress (Chemistry and Technology), 23–5 September*, Kuala Lumpur: PORIM.

Greenaway, D. and C. Milner (1993) *Trade and Industrial Policy in Developing Countries: A Manual for Policy Analysis*, London: Macmillan.

Grossman, G. and E. Helpman (1991) 'Growth and welfare in a small and open economy', in E. Helpman and A. Razin (eds) *International Trade and Trade Policy*, Cambridge, MA: MIT Press.

Helpman, E. and P. Krugman (1989) *Trade Policy and Market Structure*, Cambridge, MA: MIT Press.

Herbert-Copley, B. (1990) 'Technical change in Latin American manufacturing firms: review and synthesis', *World Development*, 18(11): 1457–69.

Jacobsson, S. (1993) 'The length of the infant industry period: evidence from the engineering industry in South Korea', *World Development*, 21(3): 407–19.

Jamil Jan (1982) 'Policy issues and implications for future development of resource-based industries in Malaysia', in M.I. Ahmad and M. Abdullah (eds) *Proceedings of the National Conference on the Development of Resource-Based Industries, August 10–12*, Serdang: Universiti Pertanian Malaysia.

Jesudason, J.V. (1989) *Ethnicity and the Economy: The State, Chinese Business, and Multinationals in Malaysia*, New York: Oxford University Press.

Johnson, C. (1982) *MITI and the Japanese Miracle*, Standford: Stanford University Press.

Jomo, K.S. (1985) *The Sun Also Sets*, 2nd edn, Kuala Lumpur: INSAN.

—— (1986) *A Question of Class*, Singapore: Oxford University Press.

—— (1990) *Growth and Structural Change in the Malaysian Economy*, London: Macmillan.

—— (1996) 'Rents in development', concept paper for conference 'Rent and Development in Southeast Asia', Kuala Lumpur, 27–8 August.

Jomo K.S. and C.B. Edwards (1993) 'Malaysian industrialisation in historical perspective', in Jomo K.S. (ed.) *Industrialising Malaysia: Policy, Performance and Prospects*, London: Routledge.

Jones, L. and I. Sakong (1980) *Government, Business, and Entrepreneurship in Economic Development: The Korean Case*, Cambridge, MA: Harvard University Press.

Kaldor, N. (1979) 'Equilibrium theory and growth theory', in M.J. Boskin (ed.) *Economics of Human Welfare: Essays in Honour of Tibor Scitovsky*, New York: Academic Press.

Kalecki, M. (1976) *Essays on Developing Economies*, Hassocks: Harvester.

Kang, D.C. (1995) 'South Korean and Taiwanese development and the new institutional economics', *International Organization*, 49(3): 555–87.

Katz, J.M. (1984) 'Domestic technological innovations and dynamic comparative advantage: further reflections on comparative case study program', *Journal of Development Economics*, 16(1–2): 13–38.

—— (ed.) (1987) *Technology Generation in Latin American Manufacturing Industries: Theory and Case Studies Concerning its Nature, Magnitude and Consequences*, London: Macmillan.

Khan, M. (1989) 'Clientelism, corruption and the capitalist state', PhD dissertation, University of Cambridge, Cambridge.

—— (1996) 'The input-output function for rent-seeking: a comparative analysis of differential effects', paper presented at a conference on 'Rent and Development in Southeast Asia', Kuala Lumpur, 27–8 August.

Khera, H.S. (1978) 'Production and consumption of oil palm products in Malaysia', Market Development of Palm Oil Products, Proceedings of a seminar organised by the International Trade Centre (ITC) UNCTAD/GATT of the United Nations, 27 March–1 April, Geneva: ITC UNCTAD/GATT.

Khong H.L. (1991) 'Service sector in Malaysia: structure and change', unpublished PhD thesis, University of Cambridge.

Khor Kok Peng (1987) *The Malaysian Economy in Decline*, Penang: Consumers Association of Penang.

Kornai, J. (1979) 'Appraisal of project appraisal', in M.J. Boskin (ed.) *Economics of Human Welfare: Essays in Honour of Tibor Scitovsky*, New York: Academic Press.

Krueger, A.O. (1974) 'The political economy of the rent-seeking society', *American Economic Review*, 64(3): 291–303.

Krugman, P. (1980) 'Scale economies, product differentiation and the pattern of trade', *American Economic Review*, 70: 750–9.

—— (1989) 'Import-protection as export promotion: international competition in the presence of oligopoly and economies of scale', in P. Krugman and E. Helpman (eds) *Trade Policy and Market Structure*, Cambridge, MA: MIT Press.

Kuala Lumpur Stock Exchange, *Annual Report*, Kuala Lumpur: KLSE, various years.

Lal, D. (1983) *The Poverty of Development Economics*, London: Hobart.

Lall, S. (1980) *The Multinational Corporation*, London: Macmillan.

—— (ed.) (1984) 'Special issue on exports of technology by newly industrialising countries', *World Development*, 12(5–6).

—— (1987) *Learning to Industrialize: The Acquisition of Technological Capability in India*, London: Macmillan.

—— (1992) 'Technological capabilities and industrialization', *World Development*, 20(2): 165–86.

—— (1995) 'Governments and industrialisation: the role of policy interventions', Panel V – Industrial Policy Reforms: The Changing Role of Government and the Private Sector Development, 'Global Forum on Industry: Perspectives for 2000 and Beyond', UNIDO, New Delhi, 16–18 October.

—— (1996) 'Technology policy and challenges', paper presented at conference, 'Globalization and Development: Lessons for the Malaysian Economy', Kuala Lumpur, August.

Landau, D.F, G.J. Schinasi, M. Cassard, V.K. Ng and M.G. Spencer (1990) 'Effects of capital flows and the domestic financial sectors in APEC developing countries', in M.S. Khan and C.M. Reinhart (eds) *Capital Flows in the APEC Region*, Occasional Paper 122, International Monetary Fund, Washington, DC.

Lazonick, W. (1996) 'Organization, finance and late industrialization', paper for conference, 'Globalization and Development: Lessons for the Malaysian Economy', Kuala Lumpur, August.

Lee, J. (1995) 'Comparative advantage in manufacturing as determinant of industrialization – the Korean case', *World Development*, 23(7): 1185–214.

Levy, B. (1994) 'Transaction costs, the size of firms and industrial policy', *Journal of Development Economics*, 34: 151–78.

Lewis, W.A. (1955) *The Theory of Economic Growth*, London: Allen and Unwin.

Little, I. (1982) *Economic Development: Theory, Policy and International Relations*, New York: Basic Books.

Lucas, R.E. (1988) 'On the mechanics of economic development', *Journal of Monetary Economics*, 22(1): 3–42.

Luedde-Neurath, R. (1986) *State Intervention and Export-oriented Development: A Reassessment of the South Korean Case*, Boulder, CO: Westview Press.

Machado, K.G. (1994) 'Proton and Malaysia's motor vehicle industry', in Jomo K.S. (ed.) *Japan and Malaysian Development: In the Shadow of the Rising Sun*, London: Routledge.

Malaysia (1968–90) *Industrial Survey*, Kuala Lumpur: Department of Statistics, various years.

Malaysia (1992) *Incentives and Guidelines for Malaysian Manufacturing*, Kuala Lumpur: Malaysian Industrial Development Authority.

Malaysian Business (1989) 'Sapura holdings: the right signals', 1–15 September: 29–33.

—— (1990a) 'Sizing up Sapura', 1–15 April: 17–18.

—— (1990b) 'Sapura spreads its wings', 1–15 April: 19–21.

—— (1992) 'In touch with market needs', 16–29 February: 45–51.

Mansor Md. Isa (1993) 'Newly emerged Bumi businesses in Malaysia: the case of Sapura Holdings', in Fujio Hara (ed.) *Formation and Restructuring of Business Groups in Malaysia*, ASEDP no. 23, Tokyo: Institute of Developing Economies.

MIDA (Malaysian Industrial Development Authority) (1988) *Statistics on the Manufacturing Sector*, Kuala Lumpur: MIDA.

MIDA (Malaysian Industrial Development Authority)/UNIDO (1985) *Medium and Long Term Industrial Master Plan Malaysia: Palm Oil Products Industry, Report*, Volume 2, Part II, Kuala Lumpur: MIDA.

MIER (Malaysian Institute of Economic Research) (1990) *Palm Oil Processing Industry: Perspectives, Challenges and Prospects* (Final Report), Kuala Lumpur: MIER.

Myrdal, G. (1957) *Economic Theory and Underdeveloped Regions*, New York: Methuen.

Nelson, R.R. (1987) 'Innovation and economic development: theoretical retrospect and prospect', in J.M. Katz (ed.) *Technology Generation in Latin American Manufacturing*

Industries: Theory and Case Studies Concerning its Nature, Magnitude and Consequences, London: Macmillan.

Ozaki, R. (1991) *Human Capitalism: The Japanese Enterprise System as World Model*, Tokyo: Kodansha International.

PDC (Penang Development Corporation) (1985–93) *Directory of Local Metal Supporting Industries in Penang*, Penang: Penang Development Corporation.

Porter, M. (1990) *The Competitive Advantage of Nations*, London: Macmillan.

Prebisch, R. (1962) 'The economic development of Latin America and its principal problems', *UN Economic Bulletin for Latin America*, February.

Proton (1995) *Corporate Profile*, Shah Alam: Proton.

Rais H. Mohamed Ariff (1995) 'Telekom Malaysia Berhad amidst changing era in the wake of liberalisation and deregulation of telecommunication sector in Malaysia', unpublished, International Islamic University, Malaysia.

Ranis, G. and J.C.H. Fei (1975) 'A model of growth and employment in the open dualistic economy: the cases of Korea and Taiwan', in F. Stewart (ed.) *Employment, Income Distribution and Development*, London: Frank Cass.

Rasiah, R. (1990) 'Export processing industries', Vol. 5, Review of the Industrial Master Plan, Report submitted to the Economic Planning Unit and UNIDO, Kuala Lumpur.

—— (1992) 'Foreign manufacturing investment in Malaysia', *Economic Bulletin for Asia and Pacific*, 63(1): 63–77.

—— (1993a) *Pembahagian Kerja Antarabangsa*, Kuala Lumpur: Malaysian Social Science Association.

—— (1993b) 'Free Trade Zones and industrial development in Malaysia', in Jomo K.S. (ed.) *Industrialising Malaysia: Policy, Performance and Prospects*, London: Routledge.

—— (1993c) 'Competition and work in the textile and garment industries in Malaysia', *Journal of Contemporary Asia*, 23(1).

—— (1993d) 'Transnational corporations and backward sourcing in the electronics industry: a study of subcontracting links with local suppliers in Malaysia', Bangkok: United Nations Economic and Social Commission for Asia and Pacific.

—— (1994a) 'Flexible production systems and local machine tool subcontracting: electronics component transnationals in Malaysia', *Cambridge Journal of Economics*, 18(3).

—— (1994b) 'AFTA and its implications for industrial restructuring: televisions and airconditioners in Malaysia', report submitted to the Thailand Development Research Institute (TDRI), Bangkok.

—— (1995) *Foreign Capital and Industrialization in Malaysia*, London: Macmillan.

—— (1996a) 'Government–business coordination and operative flexibility: electronics transnationals and local linkages', paper presented at the American Social Science Research Council Conference, Hsinchu City, 3–6 October.

—— (1996b) 'State intervention, rents and Malaysian industrialization', in John Borrego, Alejandro Alvarez Bejar and Jomo K.S. (eds) *Capital, the State, and Late Industrialization: Comparative Perspectives in the Pacific Rim*, Boulder, CO: Westview Press: 151–70.

—— (1997a) 'Political economy of Malaysia', in R. Robison, G. Rodan and K. Hewison (eds) *The Political Economy of Southeast Asia*, Sydney: Oxford University Press.

—— (1999) 'Malaysia's national innovation system', in Jomo K.S. and G. Felker (eds) *Technology, Competitiveness and the State: Malaysia's Industrial Technology Policies*, London: Routledge.

Rasiah, R. and H. Osman-Rani (1995) 'Enterprise training in Malaysia's manufacturing sector', paper presented at the World Bank Conference, 'Enterprise Training and Productivity in the Manufacturing Sector', Washington, DC, 10–13 June.

Ravenhill, J. (1996) 'Rents and development: a Schumpeterian approach', paper for conference, 'Rent and Development in Southeast Asia' Kuala Lumpur, 27–8 August.

Richardson, G. (1972) 'The organization of industry', *Economic Journal*, 82: 883–96.

Richardson, J.D. (1971) 'Constant-market-share analysis of export growth', *Journal of International Economics*, 1: 227–39.

Romer, P. (1986) 'Increasing returns and long-run growth', *Journal of Political Economy*, 94: 1002–37.

Roth, S. (1995) 'Rediscovering its own strength: lean production in the German automobile industry', European Foundation for the Improvement of Living and Working Conditions, Working Paper no. WP/95/66/EN.

Saham, J. (1980) *British Industrial Investment in Malaysia, 1963–1971*, Kuala Lumpur: Oxford University Press.

Sapura Holdings Berhad, *Annual Report*, various years.

Scherer, F. (1992) *International High Technology Competition*, Cambridge, MA: Harvard University Press.

Schumpeter, J.A. (1934) *The Theory of Economic Development*, Cambridge, MA: MIT Press.

—— (1975) *Capitalism, Socialism and Democracy*, New York: Harper and Row.

Sheperd, G. (1980) 'Policies to promote industrial development', in K. Young, W.C.F. Bussink and P. Hasan (eds) *Malaysia: Growth and Equity in a Multiracial Society*, Baltimore, MD: Johns Hopkins University Press: 182–210.

Shinohara, M. (1982) *Industrial Growth, Trade and Dynamic Patterns in the Japanese Economy*, Tokyo: University of Tokyo Press.

Soesastro, H. (ed.) (1994) *Indonesian Perspectives on APEC and Regional Cooperation in Asia Pacific*, Jakarta: CSIS.

Stewart, F. and J. James (eds) (1982) *The Economics of New Technology in Developing Countries*, London: Frances Pinter.

Suehiro, A. (1996) 'Weak state and innovative business groups – a case study of the telecommunications industry in Thailand', paper presented at conference 'Rent and Development in Southeast Asia', Kuala Lumpur, 27–8 August.

Teitel, S. and F. Thoumi (1986) 'From import substitution to exports: the manufacturing export experience of Argentina and Brazil', *Economic Development and Cultural Change*, 34(3): 455–90.

Teitel, S. and L.E. Westphal (eds) (1984) 'Symposium on technological change and industrial development', *Journal of Development Economics*, 16(1–2).

Telekom Malaysia Berhad, *Annual Report*, various years.

Thoburn, J.T. (1977) *Primary Commodity Exports and Economic Development: Theory, Evidence and a Study of Malaysia*, London: John Wiley.

Todd, R.M. (1978) 'Malaysia's experience in the export of processed palm oil', Market Development of Palm Oil Products, Proceedings of a seminar organised by the International Trade Centre (ITC) UNCTAD/GATT of the United Nations, 27 March–1 April, Geneva: ITC UNCTAD/GATT.

Towers, E. (1992) 'Domestic resource cost', *Journal of International Economic Integration*, 7: 20–44.

UNIDO (1995) 'Global trade liberalization: implications for industrial restructuring', Global Forum on Industry Perspectives for 2000 and Beyond, 16–18 October, New Delhi.

Van Duren, E., L. Martin and R. Westgren (1991) 'Assessing the competitiveness of Canada's agrifood industry', *Canadian Journal of Agricultural Economics*, 39: 727–38.

Wade, R. (1990) *Governing the Market: Economy Theory and Role of Governments in East Asian Industrialization*, Princeton: Princeton University Press.

Weiss, J. (1986) 'Industrial projects and the importance of assumptions about learning and technical change', *Project Appraisal*, 1(3): 169–76.

—— (1989) 'Assessing the technical impact of new investment projects in developing countries', *Journal of International Development*, 1(4): 496–507.

Westphal, L.E. (1981) *Empirical Justification for Infant Industry Protection*, World Bank Staff Working Paper no. 445, World Bank, Washington, DC.

Wilkinson, F. and J.I. You (1992) 'Competition and cooperation: towards an understanding of the industrial district', Small Business Research Centre (SBRC), working paper, University of Cambridge, Cambridge.

Williamson, O. (1979) 'Transaction-cost economics: the governance of contractual relations', *Journal of Law and Economics*, 22: 233–62.

—— (1985) *The Economic Institutions of Capitalism: Firms, Markets and Relational Contracting*, New York: Free Press.

World Bank (1980) *1980 World Development Report*, New York: Oxford University Press.

—— (1990) *1990 World Development Report*, New York: Oxford University Press.

—— (1993) *The East Asian Miracle: Economic Growth and Public Policy*, New York: Oxford University Press.

Yakin, F. (1991) 'Enhancing export capability of rubber-based industry', in S.H. Tan (ed.) *Enhancing Malaysia's Export Capabilities*, Kuala Lumpur: ISIS Malaysia.

Young, K., W.C.F. Bussink and P. Hasan (eds) (1980) *Malaysia: Growth and Equity in a Multiracial Society*, Baltimore, MD: Johns Hopkins University Press.

6 Economic policy and selective industrial development in Thailand

The Thai gems and jewellery industry and the Siam Cement Group

Patcharee Siroros and Wilaiwan Wannitikul with Medhi Krongkaew

The role of government and other institutional factors in selective promotion of industries and firms as part of the development process will be discussed in this chapter using an institutional framework. After critically reviewing alternative approaches and providing an analytical framework for this study, we review Thai industrial development after the Second World War. The chapter then reviews how the bureaucracy, political power and private business have been involved in the process of economic policy determination in Thailand. We then use institutional analysis to help explain the growth of particular industries, namely the Thai gems and jewellery industry and the Siam Cement Group. The post-economic crisis strategies of the two case studies are then summarised in an epilogue.

Analytical framework

In the past three decades, debate on Thai economic performance has focused on the importance of the state's role and capacity in national development from three different approaches.

The first is of modernisation theorists led by scholars such as Fred W. Riggs (1966), who characterised Thailand from 1950s to 1973 as a 'bureaucratic polity', in which all political decisions were controlled by state officials. Extra-bureaucratic forces – such as formal organisations, classes or interest groups – did not play a significant role in the policy-making or implementation processes. In this context, the businesspeople, who have mainly been ethnic Chinese, did not have political power and had to develop patron–client relationships with Thai officials to protect and advance their economic interests. Riggs explicitly rejected the possibility of substantial increases in the political power of these business entrepreneurs. Riggs speculated that if a real Sino-Thai business community could emerge, it might lay the basis for an effective transformation of the polity in the direction of a more diffracted and democratic system (though he thought this trend was unlikely). However, by the late 1990s, more than thirty years after Riggs' prediction, the economy has grown, and institutions necessary for economic growth have emerged. Businesspeople have become increasingly capable of gaining support from the state for policies favouring their interests and activities. Riggs' model provided no explanation of how this could happen.

Second, the dependency approach has tried to explain Thai development and growth. Scholars such as Grit Permtanjit (1982) described Thailand as dependent capitalist and therefore trapped in a vicious cycle of dependency on foreign aid, technical help and foreign investment. State policies and laws passed during the 1950s and 1960s ostensibly to encourage economic development were, in his view, designed to serve multinational firms and foreign interests. Grit predicted that Thailand in the 1980s would become more politically and economically anarchic, with the domestic bourgeoisie remaining weak. Grit's analysis emphasised external actors and their effects on Thailand. His analysis was similar to Riggs with regard to the presumed weakness of the Sino-Thai business community and its inability to develop into a strong force.

In contrast to the position of the modernisation and dependency approaches, the third approach by neo-Marxist scholars such as Kraisak Choonhavan (1984) and Kevin Hewison (1985) point to the emergence of a dynamic Sino-Thai bourgeoisie who have become increasingly influential and have their own interests as a class or as class fractions, rather than simply acting as compradors for multinational capital. The major contribution of these scholars has been to trace the emergence of a Sino-Thai business class and to examine its relationship to the state. Also, neo-Marxist scholars recognise that the Thai state has played an important role in the development of domestic capitalism. However, there are weaknesses in the neo-Marxist approach to understanding Thai development, especially because it tends to ignore conflict within the business class and how these are resolved.

The above three approaches which emphasise the role of the state have thus failed to explain Thai economic growth in recent decades. Institutionalist scholars – such as Doner (1991), Doner and Ramsay (1989), Anek (1989) and Patcharee (1992) – have started to explore the role of non-state institutions such as business associations, groups and networks, as well as public–private sector consultation. This approach emphasises the importance of the political organisation of business interests, and negotiations between organised business and government.

Institutionalist scholars thus argue that Thailand's success is based on the creation of flexible relations between Sino-Thai businesspeople and Thai state officials. In the past, especially between 1932 and the early 1950s, the relationship was quite tense, but in recent decades, these tensions have eased. For Sino-Thai businesspeople to prosper, they have needed and enjoyed state support. For the state to secure revenue, it has had to rely heavily on the profitable performance of these businesspeople. As the Sino-Thai business class has grown and developed since 1958, it has sought new ways to enhance its income and its relationship with the Thai state. The automobile industry is one sector which has witnessed the emergence of a business association with Sino-Thai as well as Japanese participation since the early 1970s, which Doner and Patcharee (1995) consider to be a corporatist relationship. Similar arguments have been made for the textile and sugar industries (Doner and Ramsay, 1989). Such relations are considered to be a major factor in Thai economic growth.

Although official business associations exist, informal institutions – such as Sino-Thai business groups, old classmate networks, family ties and trust – have not disappeared (Suehiro, 1996). In the automobile industry, informal institutions still play a role in negotiations and resolving conflicts within the business sector. The long-term relationships between the assemblers and suppliers, partly based on trust, have governed the prices and quality of the products (Doner and Patcharee, 1993). It is interesting to study how interrelations among state, business associations and informal institutions have developed and how their interactions affect Thai industry.

The institutional approach emphasises how institutional factors can resolve collective action problems. In the automobile and textile industries, technology development and labour shortages are some of the major problems. Industry and the Thai state work together to invest in R&D and to set up training programmes and technical training institutes to solve these types of problems (Doner and Patcharee, 1995).

Networking is another area that institutional scholars emphasise. How have networks affected the performance of industry? Doner and Ramsay (1989) found that their connection facilitated Bangkok Bank's extended credit to the Sukree textile group. Patcharee (1992) learnt that Nissan came to Thailand after the Second World War by way of a Sino-Thai businessman named Thavorn Pornprapa, who imported (completely built) Nissan trucks and later completely knocked-down kits to assemble into buses. With his links to General Praman Adireksarn, then Director-General of the Public Transportation Organisation of Thailand, Thavorn was able to sell Nissan's Datsun trucks to that organisation. Nissan also ensured special assistance, including soft loans and after-sales service, to Thavorn's Siam Motor.

How does institutional analysis assist us in understanding the 'success' of the Thai gems and jewellery industry and the Siam Cement Group? The background of the two groups show that institutional factors, such as co-operation between the Thai state and business, are major factors accounting for growth. Non-state factors, such as ties and contacts between firms, have helped solve problems of skilled workers and R&D. The Siam Cement Group is a good example of how a company benefits from ties with its associate companies in terms of training and research investment. Business associations have also played a role in solving collective problems of the Thai gems and jewellery industry in which the Thai Gems and Jewellery Traders Association has successfully lobbied for tax exemption for raw gemstones and jewellery. The Association also worked with the government to found the Gems and Jewellery Development Institute. The business community and the Ministry of Industry have developed a skilled labour training project for villagers to help overcome labour shortages for the industry.

Industrial development

Before turning to our case studies, an overview of the industrialisation process in Thailand will give some insights into Thai industrial development. Modern economic development in Thailand is said to have started after the end of the Second World War. Its First Economic Development Plan was initiated and launched in 1961, with the industrial sector assigned the role of leading sector since then. Thailand is now under its ninth five-year Economic and Social Development Plan. During three decades of development, the Thai economy has changed from an agricultural-based economy to the point of becoming a newly industrialised economy.

As the leading sector of the economy, manufacturing industry has become the focal point of the development process. Industrial development strategies have been proposed in the economic and social development plans. Thailand, like other developing countries, started industrial development by import substitution as the local market was ready to absorb all domestic production of previously imported goods. However, once the domestic market became saturated, the next stage involved a choice: (a) domestic production of previously imported intermediate, capital and more durable consumer goods; (b) an export-oriented manufacturing; or (c) a combination of these two.

Thailand followed an import-substitution strategy in the first decade of its development plans (1960–70). In addition, a policy to encourage private investment was adopted after a largely unsuccessful attempt to promote manufacturing investment through public enterprises during the mid-1950s. The Board of Investment (BOI) was established in 1959 to carry out private investment promotion, and a combination of incentive schemes, tariff policies, tax regimes, trade and price controls gave direction to the pattern of industrial investment (World Bank, 1980: 13). The policy of import substitution during this decade was reflected in the structure of import tariffs, which were raised significantly in 1964 and again in 1970 to provide protection for domestic industries. Towards the end of the decade, while investment incentives were being reduced, the level of tariff protection increased. As the limits of import substitution in a relatively small domestic market were reached in the early 1970s, Thailand embarked upon an export-oriented policy of labour-intensive manufacturing.

By 1971, there was a significant bias favouring production for the domestic market and opposing exports, with incentives being strongest for production of final products based on imported intermediates and capital goods (World Bank, 1980: 13–15). In 1972, exports were actively promoted by the Export Promotion Act and setting up of the Export Promotion Committee to co-ordinate export-enhancing measures such as full tax exemption on imported machinery and raw materials; exemptions from certain business taxes; refunds of all taxes incurred for the production process; a rediscount credit facility at subsidised interest rates from the Bank of Thailand (BOT); and technical assistance from the Export Service Centre of the Ministry of Commerce. 'These promotion measures contributed to the phenomenal growth of manufacturing exports between 1970 and 1976 at an annual compound rate of about 30 per cent' (World Bank, 1980: 14).

However, industrial policy and investment promotion in the 1960s and 1970s (the first four plans) led to uneven industrial location. Approximately 85 per cent of industrial establishments were concentrated in the greater Bangkok metropolitan area. Thus, promotion of industries outside the Bangkok area became one of the major objectives of the government's industrial policy in the Fifth National Development Plan. Measures to promote industries in rural areas included a 50 per cent deduction in corporate income tax for five years following the first eight years of full exemption available to all promoted firms, and up to 90 per cent exemption for five years from business taxes, tax deduction of twice the actual cost of transport, water and electricity, and credit assistance by the Bank of Thailand and the Industrial Finance Corporation of Thailand (IFCT) were also available. Unfortunately, these incentives appear to have had little impact on the location of industries.

For the rest of the 1980s and in the 1990s, three major groups of industries were targeted:

1 export-oriented industries;
2 small and rural industries located in areas outside Bangkok and its vicinity; and
3 development of basic industries, i.e. engineering, petrochemical and iron and steel industries.

Thailand is now under its Ninth Economic and Social Development Plan (2002–6); its industrial development has progressed in the past thirty years through several phases, from an originally small manufacturing sector, mainly consisting of industrial processing of primary products, to import substitution of consumer goods for the domestic market in 1960–70, and to an emphasis on export-promotion of labour-intensive industries in the 1970s. In the mid-1980s, Thailand's manufacturing sector still had most of the characteristics of a predominantly agricultural, semi-industrialised economy. A large part of manufacturing still consists of cash-crop processing, much of it by small-scale enterprises scattered throughout the country. Domestic production of capital goods is still very small or even negligible. Nevertheless, in the second half of the 1980s, Thailand was overheating due to fast growth. The influx of foreign direct investment pushed the industrial sector towards more export-orientation and greater import substitution of intermediate products. By 1997, Thailand could claim to be an emerging newly industrialising country (NIC).

Political transformation

When Thailand changed from an absolute monarchy into a constitutional monarchy in 1932, the country was run by military officers and civilian bureaucrats who had brought about the political change. This system of government was known as a 'bureaucratic polity' – a term coined by Fred W. Riggs, an American public administration specialist – because although it had an appearance of democracy, real political power was still concentrated within a small group of military leaders and civilian bureaucrats, including both decision-makers and policy implementers.

Under this politically dominant military and civilian bureaucracy, private business and other social forces were practically left out. As most private businesses in Thailand were run by ethnic Chinese business entrepreneurs, social discrimination against the Chinese by the government at the time politically weakened this Chinese commercial class. However, these Chinese businesspeople survived, partly by associating themselves with the more influential Western business community operating in Thailand, or by 'continually buying protection from the Thai elite' (Riggs, 1966: 251). They were what Riggs called 'pariah entrepreneurs', whose wealth and income from business success had to be used to buy protection from government elites and bureaucrats. The post-war government attempted to check the economic influences of Western as well as ethnic Chinese business firms by setting up numerous state enterprises. Only when this attempt failed largely because of government inefficiency in running these state enterprises did the government change its policy to accommodate the Chinese business community.

However, the toppling of the military government in 1973 by a student-led political uprising exposed the vulnerability of this bureaucratic form of governance. Political scientists such as Hewison (1989) began to question the relevance and applicability of this model to Thailand. To him, in the latter part of the 1970s, Thailand had changed politically, socially and economically. The power of the state no longer remained monolithic in the hands of the military-civilian bureaucracy. The rise of the Sino-Thai capitalist class in the post-war period proved that it was not merely a comprador class depending on foreign capital for its development, but an independent and autonomous capitalist class. The economic strength of the capitalist class redefined the relationship between the state and business in such a way that the state 'has developed policies, strategies and rules beneficial to the capitalist class (or particular fractions of it), and has protected the property interests of this class' (Hewison, 1989: 2).

Several other political scientists joined in the rejection of the bureaucratic polity as the relevant political model for Thailand in the post-1970s period in favour of a bureaucratic-authoritarian model. Patcharee (1985) argued that the bureaucratic polity (along with pluralistic and early dependency approaches) failed to explain the Thai state's relationship with business groups from the 1960s. She used cases of the automotive and electrical appliances industries, where the influences of foreign multinational companies (which were in control of these two industries) intertwined with those of state bureaucrats and local business to shape policies for these two industries. Ramsey (1987) did not reject the bureaucratic polity model outright, but used a case study of sugar policy in Thailand during the 1970s and 1980s, where the rise of the (private) sugar farmers' associations and sugar millers' associations played an important role in the making of Thailand's sugar policy to suggest modification of the traditional model.

Perhaps the strongest and most recent criticism of the bureaucratic polity notion has come from Anek Laothamatas (1992a). Anek contended that the dramatic economic development of the past three decades has spawned extra-bureaucratic forces that have weakened the hold of the military-civilian bureaucracy. Chief among these extra-bureaucratic forces has been the greater role of business associations, both in Bangkok and the provinces. Anek pointed out that since the late 1970s, organised business has suggested, initiated, amended, or blocked important economic policy and government legislation. The business associations have been successful in proposing a more export-oriented industrial development strategy, tax and tariff cuts, reductions of bureaucratic delays and inefficiency, and so on. Viewed in this light, Anek argued that the Thai polity could hardly be termed a bureaucratic polity since at least major economic decision-making had ceased to be the monopoly of the bureaucracy, but was increasingly influenced significantly by organised business. To him, businesspeople could now influence public policy through their strategic roles in the increasingly capitalist economy, or through their clientelistic ties with high officials, or simply through organised pressure groups.[1]

Although Anek asserted that the Thai bureaucracy no longer monopolised economic policy-making, it is erroneous to conclude that Thai bureaucrats no longer had influence and power over the making of economic policy. Rangsun Thanapornpun (1989: 146) had argued that it would be too hasty to make such a conclusion considering the evidence from 1973 to August 1988 (the period covered by Rangsun's study). Although democratic forces had a larger role in making economic policy, the power elites, drawn from the bureaucracy and technocrats, were still very influential.

This was not to argue that the traditional bureaucratic polity had remained unchanged since the 1960s. Like many other systems of governance, both time and circumstances had made impacts. For example, Chai-Anan (1988), a well-known proponent of the bureaucratic polity notion, admitted that Riggs' original concept of a bureaucratic polity had to be modified to incorporate changes in the social, economic and political environments. He was willing to recognise and incorporate the emergence of other societal forces – such as private business, other non-government organisations, or extra-bureaucratic groups – alongside the powerful bureaucracy. As Sakkarin (1995: 23) has correctly observed, Chai-Anan believed in the capability of the Thai bureaucracy to adjust itself to maintain a leading role in society in spite of the rise of the private sector.

Sakkarin (1995: 24) himself, however, believed that state–society relations in Thailand have been marked by 'shifting patterns of interdependence', but did not accept that non-bureaucratic interests necessarily needed support from the state to expand and compete. Although he agreed with Hawes and Liu (1993) that both the state and societal forces find themselves in shifting patterns of interdependence, for him 'shifting patterns' were not simply dictated by a 'growth coalition' sharing the desire for economic growth, but rather by a more complex coalition of 'pro-reform' and 'anti-reform' interests, comprising elements of the bureaucracy collaborating and/or competing with business and other non-state actors. In other words, bureaucratic and non-bureaucratic interests can be either strategic partners or hostile rivals, depending on the policy issue or the political regime (Sakkarin, 1995: 24).[2]

To conclude, Thailand in the 1990s is very different from Thailand in the 1950s or 1960s, when the bureaucratic polity undoubtedly governed. During the past two to three decades, several other societal forces have emerged and been strengthened, so much so that no one force could claim to dominate all other forces in influencing or determining economic policy. Contrary to the claims about the demise of bureaucratic power in Thailand and its eclipsed role in policy determination, there is considerable evidence of the bureaucracy's continued influence and power.[3]

Economic policy-making by a business-oriented bureaucracy

Despite changes in the political structure with rapid economic development in Thailand in the past two decades, the influence of Thai bureaucrats in economic policies remains strong. It is undeniable, of course, that the rapid economic growth and industrialisation of the Thai economy has brought about changes in the political structure. The private business class in manufacturing, trade and services, and individuals or groups in this class have made significant in-roads into economic policy-making in various ways. But in the final analysis, bureaucrats and technocrats in the public sector still hold the key to the formulation and implementation of most economic policies.

Obviously, the rise of the business class, the 'new middle class' in Thai society, poses a new challenge to the bureaucrats from among the 'traditional middle class'. The income and wealth associated with the new business class have put pressure on bureaucrats to re-examine the goals and means of development, whether to adopt more socially-inclined development policies or more business-oriented policies stressing efficiency and growth. The erosion of welfare arrangements associated with working in the public sector has caused a serious depletion in the resources of the Thai bureaucracy, both in terms of quantity and quality. A weakened bureaucracy is more prone to being influenced and overshadowed by business and other interest groups in society, which further accelerate the reorientation of economic policy. However, as the administrative structure of the Thai state has deep roots in society, undisturbed by foreign colonisation, the environment in which the Thai bureaucracy operates still has considerable power.[4]

There are at least three observations that can be made in conclusion about the important institutional changes in economic decision-making. First, Thai bureaucrats are not a homogeneous group of people with the same orientation and outlook. They may be bound by the same administrative duties to work for the state and to protect the interests of the state, but their specific responsibilities and powers could cause them to come into conflict with one another. Christensen *et al.* (1992) suggested a bifurcation between macro and micro economic policies in the Thai bureaucracy. On the one hand, there are bureaucrats or technocrats who specialise on broader macroeconomic issues, e.g. those in the Bank of Thailand, the National Economic and Social Development Board (NESDB) and the Ministry of Finance, who often prefer conservative macro policies based on stability and growth.

On the other hand, there are bureaucrats in other ministries who are more concerned with microeconomic or sectoral policies with narrower objectives.

Christensen *et al.* (1992: 50–1) claimed that the macroeconomic technocrats typically distrusted the line ministries, believing the latter were dominated by narrow bureaucratic interests. In these ministries, the promotion of private firms was likely to involve rent-seeking, rather than socially-desirable objectives. There were tensions – rather than co-ordination and complementarity – in the relationship between these two sets of policies. They further claimed that the bifurcation of policy allowed a stable macroeconomic policy regime, but when systematic attention to a particular problem was required, overall performance was less effective. These adversarial arrangements and the lack of formal linkage mechanisms between macro and sectoral policy officials often limited the state's ability to formulate and co-ordinate overall development policy (Christensen, 1992: 52).

Second, the experience of economic policy-making under Anand Panyarachun was quite unique and opened up the possibility of more independent technocratic economic policy-making. Anand himself and most of his cabinet members were neither elected politicians nor coup leaders, but technocrats given additional powers. In this regard, the Anand government saw itself as a technocratic government, rather than a government of business.

Third, the business-oriented bureaucracy, in co-operation with other non-bureaucratic sectors in the economy, has succeeded in fostering an open, competitive, growth-oriented development strategy. Before 1996, the rate of GDP growth in the previous two decades had averaged about 7 per cent per annum in real terms. This was very high by international standards, and resulted in the per capita GNP of Thailand rising from about US$435 in 1970 to about US$1,650 in 1992. This growth helped reduce the incidence of poverty in the whole country from about 32 per cent in 1975/76 to about 22 per cent in 1988 (Medhi *et al.*, 1992). However, this growth-oriented development policy also resulted in increased income inequality. The Gini coefficient for income distribution in Thailand increased from about 0.426 in 1975/76 to about 0.479 in 1988 and about 0.525 in 1994 (Medhi *et al.*, 1992; Kakwani and Medhi, 1997).

Fourth, unlike some other East Asian economies whose economic success has been due to industrial policy, Thailand never really had an explicit industrial policy before November 1996, when the Industrial Master Plan for Thailand was adopted by the government. This was the work of academicians in co-operation with technocrats-bureaucrats in the Ministry of Industry. Perhaps even more important has been the Industrial Restructuring Plan, which aims at pushing Thai industry from low value-added to high value-added industrial production, another important contribution from the technocrat-bureaucrats in the Thai government (see Ministry of Industry, 1996 and 1997).

The big challenge for the near future will be the capability of the Thai bureaucracy to initiate changes in economic policies to cope with economic problems in an increasingly globalised world, especially in the aftermath of the 1997 economic crisis, when Thailand was forced to open up and liberalise even more than ever. Also, there are problems that need the attention of a strong public sector, such as the environment, education, health, public safety and order. These are challenges that the present business-oriented bureaucracy may be ill-equipped (quantitatively as well as qualitatively) to respond to.

Hence, contrary to what many people think, the future economic development of Thailand needs a stronger and more capable and accountable bureaucracy. Such a bureaucracy must be able to attract the same high calibre personnel as those who have joined the private sector recently. The relevant and appropriate thing to do is not to engage in bureaucrat-bashing, or downgrading the role and contribution of bureaucracy, but rather to equip the bureaucracy with an ability to operate in the constantly changing economic situation without losing sight of the proper role of the state.

The gems and jewellery industry

Evolution and development

Thailand is one of the five largest sources of gemstones in the world, together with South Africa, South America, Myanmar and Sri Lanka. Within Thailand, the three largest sources of gemstones are in the provinces of Trat, Chanthaburi and Kanchanaburi. These three provinces account for 80 to 90 per cent of gemstone mining in Thailand. Other minor sources of gemstones include the provinces of Prae, Si Sa Ket and Sukhothai. The major types of gemstones found in Thailand are rubies and blue sapphires. Minor gemstones found in the same areas include topazes, zircons, red garnets and black sapphires.

Small-scale gemstone mining, gem-cutting and jewellery-making in Thailand began hundreds of years ago. The first gem mine, 'Bor Kaew', was in Boe Chong sub-district, Den Chai district, Prae province in northern Thailand. Most of the gemstones produced by the mine were sapphires. Later, mines were opened in Tambon Bangkaja and Kao Ploy Vaen in Chantaburi, and in Trat and Kanchanaburi provinces. Blue sapphires, rubies, yellow sapphires and star sapphires were found in these mines. In addition, brown zircons were found at a mine in Si Sa Ket province.

Historically, the art of gem-cutting began with the Thai Yai tribespeople who live in the mountainous regions in the north of Thailand. By the Second World War, the gemstone industry had begun to develop on a larger scale (Nilphechara, 1982), with the gem-cutting industry centred around Talad Noi (Small Market), Charoenkrung Road, Bangkok (NESDB, 1984). Both cutting and polishing have been done by lapidaries in Talad Noi.

The development of the gems and jewellery industry is quite fascinating. The success story of this industry is evidence of private sector cohesion, in co-operation with the public sector. This industry started with small-scale and/or home-based gem mining and gem-cutting scattered throughout the country. Being an insignificant foreign exchange earner before the 1970s, the government gave no attention to the gems and jewellery industry. Nevertheless, the emergence of Thailand as a centre for coloured gemstone trading and cutting was initially due to its abundant resources of precious and semi-precious raw materials coupled with its highly skilled gem-cutters (*Thailand Business*, July–August 1987). The country's world-renowned coloured gemstones are sapphires and rubies, which account for 70 to 80 per cent of total cut gemstone exports.

The assiduous efforts of its business associations, business groups and networks have convinced the government to recognise the fact that the gems and jewellery industry was not only a growing foreign exchange earner, but also provided gainful seasonal employment for nearly half a million farmers during their off-season. Therefore, starting from the mid-1970s, the government began seriously to promote the development of this industry. During the 1980s, especially in the second half, the industry experienced rapid growth and became one of the top five export items. The value of exports jumped from 8,518 million baht in 1985 to 34,877 million baht in 1990 – a more than quadruple increase (see Table 6.1).

The brilliant development of the industry during the 1980s suggested optimistic prospects for pushing Thailand towards becoming a world centre for gems and jewellery. However, the first half of the 1990s saw slow growth in this industry as the rate of growth of its export value declined. There was fear that the industry was losing its comparative advantage, and there was doubt whether Thailand could become the centre of the world gems and jewellery industry. Several studies were carried out to assess the prospects of this industry, and the results suggested

Table 6.1 Thailand: export value of gems and jewellery, 1981–2000

Year	*Export value (million baht)*	*Growth rate (%)*
1981	5,022	
1982	5,756	12.7
1983	8,234	30.1
1984	7,382	–1.1
1985	8,518	13.3
1986	13,164	35.3
1987	19,811	33.5
1988	23,726	16.5
1989	28,082	15.5
1990	34,877	19.5
1991	35,988	3.1
1992	36,616	1.7
1993	41,050	12.1
1994	44,685	8.1
1995	50,179	10.9
1996	54,273	7.5
1997	55,622	2.4
1998	57,350	3.0
1999	59,821	4.1
2000 (Jan.–June)	29,342	

Sources: Department of Customs (1981), cited in Bank of Thailand, 1996: 11–24 (export value of gems include artificial gems); Department of Customs (1982–8), cited in IMC, 1991: 3; Department of Business Economics (1991–4), cited in ICT, 1996: 284; Department of Business Economics (1995–6), cited in DEP, 1996: 2.

that there were still possibilities of developing it into a more sustainable one, and maintaining its competitive edge – if both the government and the private sector would join hands in solving problems and overcoming obstacles in the industry.

At present, the gems and jewellery industry is among the top ten foreign exchange earners in the country. Altogether, it involves more than a million permanent and seasonal workers, of whom about 300,000 are miners with the rest engaged in gem-cutting and jewellery manufacturing. Of overall employment, only about 25 per cent is permanent while the rest is seasonal.

Based on statistics for the development of the industry and export value, the evolution of the industry can be divided by decade into three periods: the 1970s; the 1980s; and the 1990s.

The 1970s

Before 1970, the government rather neglected this industry, assuming that it served only the rich or the upper middle class, and had little to contribute to the economy. At that time, most gem production was small-scale, with home production in gem-cutting and jewellery-making accounting for much of the industry. However, during this period, the raw material resources of the country were still abundant and the labour costs quite low. Most of the workers (about 75 per cent) in the industry were migrating poor farmers who mined for gemstones in their off-season and carried out gem-cutting activities in their homes as off-farm employment (*Thailand Business*, 1987: 35). This low labour cost is a major characteristic of this industry. Therefore, the period before the 1970s saw the steady growth of the industry without any assistance from the government, mainly due to its comparative advantage as a resource-based industry.

By the end of the 1970s, Thailand's traditional mining areas were becoming exhausted, while other mining areas were declared off-limits because they were classified as either militarily sensitive areas, forest reserve areas, or security areas. Imports of rough gemstones from other countries – such as Myanmar, Sri Lanka, Australia, Belgium, India and Pakistan – became necessary for thousands of gem-cutters in the country. The tax system of 5 per cent import duty, 10 per cent municipal tax and 3 per cent business tax increased costs, reducing competitiveness. The Jeweller's Association, the first organisation of the gems and jewellery traders, was not successful in seeking government assistance.

In 1976, a new generation of industry businesspeople founded their own association called the Thai Gems and Jewellery Traders Association (TGJTA). Led by Anand Salwalla, a prominent figure in the industry, they fought to get the government to favour their industry. They tried hard to convince the government of the important role of the gems and jewellery industry as a significant export earner and job creator. In 1977, a year later, the TGJTA succeeded in getting the government to grant promotion privileges to the industry, and to lift import duties and business taxes on rough gemstones, opening a new phase in the gems and jewellery industry's history. Medium and large-scale (employing more than

100 workers) firms started to emerge. The industry, which previously concentrated on gem-cutting and mostly exported loose cut gems, shifted to more value-added jewellery products. The government and private sector organisations co-operated to make the industry more export-oriented. Since then, the export values have steadily increased (*Phu Nam Turakit* (Business Leaders), June 1989). (Unfortunately, export statistics of the 1970s are not available.)

Towards the end of the 1970s, rough gemstones were imported to supplement local raw materials. One innovation that the Thai gemstone industry pioneered was to heat relatively worthless milky white geudas to turn them into highly-valued blue sapphires. Thus, milky white geudas were cheaply imported from Sri Lanka to make high-priced products. This caused conflicts between Sri Lanka and Thailand, resulting in a ban of Sri Lankan geuda exports. To solve the conflict, the TGJTA leaders sought assistance from the Ministry of Foreign Affairs. Anand and the TGJTA secretariat joined the Ministry of Foreign Affairs in negotiating with the Sri Lankan Ministry of Finance. After more than two years, an agreement was reached which allowed Thai gem traders holding special passports to import geudas without limitation. Thus, close co-operation between the public and private sectors solved this problem of the industry.

The 1980s

The 1980s were a fast growth period for the industry. Starting from 1980, import and business taxes on cut gemstones and diamonds were waived, and in 1981, the business tax on finished jewellery manufactured for export was also lifted. These two measures increased the competitive edge of the jewellery manufactured in Thailand, and also paved the way for diamond-cutting factories. Tax liberalisation coupled with promotional privileges for jewellery manufactured for export, translated into a further increase in export value and further expansion of the industry. The rapid increase in export value was greatest in 1986 and 1987 (Table 6.1). In 1986, industry exports were 13,164 million baht, increasing to 19,811 million baht in 1987, when it became the fifth-largest industry in terms of Thailand's exports (IMC, 1991). In 1989, exports increased to 28,082 million baht, becoming Thailand's third most important national export commodity. In 1990, the export value of the industry rose to 34,877 million baht, placing it second.

The factors contributing to the strong expansion of the industry in this period include booming international demand, especially from the United States, Japan, Hong Kong and Europe. High demand for imports of Thai gems and jewellery was due to the competitive prices of Thai products and their world quality standards. One important factor was the co-operation of the private and public sectors in promoting the industry. The government provided substantial support for Thai gems and jewellery in international markets. One of the most successful co-operative promotion measures was the organisation of the Bangkok Gems and Jewellery Fair, first started in 1985 and co-organised by the Department of Export Promotion (DEP) and the Ministry of Commerce, along with the TGJTA and the Jewellers' Association. The Fair was very successful in exposing Thai products to the international gems and jewellery trade. Consequently, the Fair has been held every year since then, and its frequency has increased to twice a year since 1991. Besides the Gems and Jewellery Fair, a promotion fund for incoming and outgoing trade missions was also set up. Other promotion activities have included international trade fairs, international joint promotion efforts, information stands, publications and image promotion. This has made the Thai gems and jewellery industry world renowned.

The fast growth period was due to substantial private investment in the industry in response to high demand for gem exports. High demand, together with the rising prices of imported and locally mined gems, led to increasing costs in the industry. Some gem-cutting factories shifted to diamond-cutting to avoid the scarcity of gemstones. The escalation of labour costs, coupled with the shortage of skilled labour, undermined the competitive position of the industry. Hence, in 1989, the government agreed to set up the Gems and Jewellery Industry Development Institute, which was affiliated with the Industrial Promotion Department, Ministry of Industry, as the key agency responsible for promoting and co-ordinating with other public and private organisations in planning and problem-solving for the industry. Moreover, the Institute acts as a service centre for analysing, evaluating and issuing guarantees for gems and jewellery products, and also as a centre for developing highly-skilled workers and new production technology. This Institute operates under guidelines set by the Gems and Jewellery Development Committee, where representatives from the gems and jewellery associations are included. The Institute began operations in 1991, with its first task that of training and upgrading the skills and craftsmanship of workers in the industry.

The 1990s

During the first half of the 1990s, the growth rate of exports in the gems and jewellery industry experienced strong decline, but the value of total exports continued to increase. In 1991 and 1992, export values expanded at 3.1 and 1.7 per cent respectively, rebounded to 12.1 per cent in 1993, before falling back. In 1994, the export value of gems and jewellery from Thailand was 44,685 million baht, i.e. a growth rate of 8.1 per cent. By 1995, export value had increased to 50,179 million baht, i.e. at a growth rate of 10.9 per cent. Finally, export value from January to June 1996 was estimated at 25,215 baht, i.e. at a projected annual growth rate of 10.5 per cent (DEP, 1996: 3). However, the gems and jewellery industry in Thailand grew more slowly in the late 1990s due to both international and domestic factors, as outlined below.

Success factors

The Thai gems and jewellery industry is one of the success stories of indigenous firms successfully improving their international competitiveness through co-operation and with government support. Factors affecting the success of the industry can be summed up as follows.

Raw material abundance and cheap labour

In the early stages of the industry's development, abundant raw material resources and cheap labour were the main factors that helped make the industry competitive. During that period, exports were mostly in the form of loose gemstones which did not require high technology. The only requirement was the skilled craftsmanship then available at lower wages. However, when the raw material base had been exploited and labour was no longer cheap, product variety and differentiation became the strategy for success.

Product variety and differentiation

There is ample scope for product variety and differentiation in the gems and jewellery industry. Different designs bring about product variety, as well as various kinds of gemstone. Different raw material quality and cutting skill differentiate products. Thus, the gems and jewellery industry could supply consumers in every market segment. When producers lost comparative advantage in the lower-end market, in order to survive they moved to the more value-added, design-intensive and higher quality production segment. The upgrading of product design and quality are, therefore, what the firms strive for.

Strong business community organisations

The business community has worked with the government to promote the gems and jewellery industry. At present, the following four business groups are active in the promotion process.

1 *The Thai Gems and Jewellery Traders Association* (*TGJTA*) is the most important private sector group. The TGJTA was founded in December 1976 to promote the export of Thai jewellery, to negotiate with the government and other international institutions to protect the interests of its members, to exchange information among members, and to develop technology and standardised products in co-operation with the government. Significant achievements of the association in the past have included:

   - proposals in 1977 and 1980 that the government exempt raw gemstones and jewellery from tariff duties and business taxes;
   - a proposal in 1981 that the government exempt ornaments for export from business and municipal taxes;
   - a proposal that the government establish the Gems and Jewellery Development Institute as the main agency to train Thai gems technicians;
   - organisation of the Bangkok Gems and Jewellery Fairs in co-operation with the Department of Export Promotion and the Jewellers' Association; and
   - negotiation with market representatives from the European Union and the United States, with the results reported to the government.

2 *The Jewellers' Association* was established in 1952 to promote gemstones and ornaments for the domestic market. The Association worked co-operatively with the TGJTA to organise seminars on the industry with Belgian business groups.

3 *The Thai Diamond Manufacturers Association* was set up in 1989 to represent foreign jewellery companies in talks with the Thai government. Its 31 members are from the foreign and joint venture companies in Thailand. Past TDMA achievements include:

- representing the Thai government in negotiations with De Beers, the world gems and jewellery organisation, in March 1991 to secure recognition of Thailand as the sixth world gems and jewellery centre;
- representing Thailand in the World Diamond Congress Meeting in May 1991. After the meeting, Thailand became a member of the prestigious International Diamond Manufacturers Association (IDMA);
- supporting the establishment in 1990 of the Jewellery Designers' Association of Thailand.

4 *The Asian Institute of Gemological Sciences* was set up in 1978 as the first of its kind in Southeast Asia to offer a course on gemological sciences for interested people who want a career in the field.

Government–private sector co-operation

The gems and jewellery industry is a success story of close co-operation between the government and private institutions to promote the industry in the past two decades. Some samples of such co-operation, besides the Bangkok Gems and Jewellery Fairs, are as follows:

- a pilot project for co-ordination of industrial investment in provinces such as Ubon Ratchathani and Si Sa Ket in the north-eastern part of the country. This project was set up in 1988 in co-operation with the Joint Public–Private Co-operation Committee (JPPCC) and TGJTA to promote production, processing and marketing as well as information exchange among entrepreneurs;
- trade delegations from the public and private sectors to find new international markets;
- establishment of the Gems and Jewellery Development Institute;
- establishment of a gems and jewellery training programme for prisoners in co-operation with the Corrections Department and the Calibration of Gems Factory, Ltd, started in 1990. Another company, Quality Color, Ltd had a training programme for 1,800 prisoners so they could work with the company after release;
- establishment of an undergraduate degree programme in gemological sciences at the Sri Nakarin Wirote University to produce graduates in the field,

involving co-operation between the Ministry of University Affairs and the industry to offer courses in the Materials Science Department, Faculty of Science, starting in 1992.

The business community and industry associations have played a significant role in working with the government to promote growth, solve problems, and set the direction for the gems and jewellery industry. Such co-operation provided an effective way to solve some collective problems in the industry involving marketing information, high production costs caused by the government's value-added tax, shortage of skilled labour, and so on. The case of the gems and jewellery industry shows the effectiveness of institutional strategies, such as setting up committees and organisations, e.g. the Gems and Jewellery Development Institute, to address the labour shortage problem, or inviting committees with members from the public and private sectors to exchange information on export markets.

However, such co-operation is lacking in the development of research and design capabilities as both sides did not see it as crucial for industrial growth, since income mainly came from selling semi-finished products in international markets. The industry's comparative advantage has thus begun to decline recently. Thai policy-makers and business people agree that design technology development is crucial for Thai brand names to succeed in international markets. The government has started a design-contest project which does not seem to have been successful apparently because the rewards are too small to attract entries. Thailand needs a new strategy and policy instrument to solve this collective action problem.

International markets

At present, diamonds, rubies, blue sapphires and emeralds are the major gemstones in the global gems and jewellery market. Thailand's major export markets for gems and jewellery are the United States, Europe (mainly Belgium, Germany, France and the United Kingdom), Japan, Hong Kong and Israel. Thailand's market shares in the United States, Europe and Japan are 5, 3 and 9 per cent respectively (ICT, 1996: 279). Currently, Thailand faces great competition in the international market for gems and jewellery, especially from India and Sri Lanka, and has largely lost its market share of low-priced, low-quality gems and jewellery to India. Thailand's exports also face the threat of increased taxes on imports of gems and jewellery into the United States, or even an import embargo to ensure Thailand's compliance with US copyright laws (BOT, 1992: 225). Other countries which import gems and jewellery ffom Thailand might also follow the US lead, increasing their own countries' non-tariff trade barriers to gems and jewellery imports from Thailand (BOT, 1993: 11–20).

Local factors

A shortage of raw gems in Thailand has continued to be a constraint on the gems and jewellery industry in Thailand. As a result, Thai entrepreneurs have made substantial investments in other regional sources of gems, including mines in Myanmar, Vietnam and Cambodia. The Thai government has imposed a value-added tax on all traded commodities since 1 January 1992 (BOT, 1992: 224). The value-added tax (VAT) has increased production costs in the gems and jewellery industry. Even though firms can obtain tax rebates, it involves a cumbersome and lengthy process to do so, and adds to the financial pressures on the gems and jewellery industry. Import taxes on raw materials used in the gems and jewellery industry such as pearls, silver, jewellery parts and equipment are very high (15 to 60 per cent of import prices), increasing production costs in the industry (BOT, 1993).

To avoid the imposition of value-added tax, a tax-free industrial estate for gems and jewellery was proposed by a group of private companies. A group of leading exporters have co-operated to set up a centre for the gems and jewellery industry called Gemopolis. The Gemopolis project comprises the Gemopolis Industrial Estate, Bangkok Diamonds and Precious Stones Exchange, Gemopolis factory village, Gemo town, Gemopolis factory outlet, duty-free shop, hotel, recreation and medical centres. The Gemopolis project has been supported by the BOI, the Industrial Estate Authority of Thailand and the Customs Department. With this support, a bonded warehouse will be set up, which will make Gemopolis a tax-free zone. The project is aimed at inducing more foreign direct investments in the industry and making Thailand a world centre for the gems and jewellery industry. Besides Gemopolis, similar projects have been proposed by other companies. Furthermore, import taxes on raw materials used in jewellery manufacturing were reduced from 15–60 per cent to 5–10 per cent in 1994, in order to enhance the competitive position of the industry.

Increasing wages for labour has increased production costs in the gems and jewellery industry. As a result, high costs of production have decreased the industry's potential to compete in international markets. A shortage of professional designers, skilled labour and managers in the gems and jewellery industry has also created production problems and decreased the industry's ability to compete abroad. Escalating wage costs and the shortage of skilled workers have put the industry in a vulnerable situation. To counteract the increase in labour costs, jewellery manufacturers have been gearing themselves to compete on the basis of product quality and have increasingly moved more into medium- and high-end products (*Bangkok Gems & Jewellery*, October 1995: 49). In addition, the private and public sectors have joined hands to help alleviate the problems.

Training and educational programmes have been proposed in both private and government academic institutions. For example, the TGJTA has been instrumental in setting up a five-year degree course in gemology at the Sri Nakarin Wirote University, where 160 students will graduate annually. The TGJTA also joined hands with the Rajamangala Institute of Technology to provide training in advanced mould production. The Department of Export Promotion (DEP) has provided a sum of 4 million baht to purchase equipment needed for the training programme which commenced in 1994. The first batch of 120 students will, upon graduation, form the nucleus of a highly-skilled workforce that will help the rapid development and growth of Thailand's gem and jewellery industry. Beside the business associations, private companies have also collaborated with the Department of Vocational Education in providing training in jewellery manufacturing techniques at Rajasthiram Technical College. These training and educational programmes for human resource development are aimed at overcoming the shortage of skilled labour in the industry.

The Siam Cement Group

The Siam Cement Group is one of Thailand's leading industrial conglomerates. The Group is known as the largest cement and construction materials producer in Thailand with a sales volume in 1996 over US$4,000 million. Siam Cement is one of the oldest and largest firms in Thailand with connections with the royal family and key government officials and recognised for the consistently high quality of its products. One factor in the success of Siam Cement has been government protection for import-substitution industrialisation in the 1960s. Therefore, Siam Cement is an extremely interesting case of how a well-established firm with a conservative management style can adjust to survive and maintain growth and success in the face of liberalisation.

Company history

Siam Cement started operations in 1915 under King Rama VI, who ruled the country from 1910 to 1925. The king was educated in England and had seen modern buildings built from concrete in that country which encouraged him to think of cement buildings and roads in Siam in future. Siam had started physical development of infrastructure during the reign of King Rama V in the face of Western colonialism. Modern roads and bridges were built to show that Thailand was a civilised nation. When King Rama VI became king, modern infrastructure projects – such as roads and government buildings – were constructed. Demand for cement increased rapidly. King Rama VI realised the significance of cement and ordered his people to start the venture in 1915. To ensure that the company would be Thai, the Crown Property Office owned half the company's shares, while the rest were owned by several nobles who worked closely with the king. At that time, Thailand lacked management and production technology, and therefore invited foreigners to join the new company. However, the founding group was afraid the company would be dominated by foreigners and issued a company regulation requiring that three-quarters of the shares had to be owned by Thais, which has been enforced ever since.

Siam Cement's growth and the government's cement policy

The growth of Siam Cement can be divided into four phases namely: (1) from 1915 to 1959; (2) from 1960, when Thailand started its First National Economic Plan, up to 1973; (3) from 1973, when Siam Cement had to face the first oil crisis, up to 1989; (4) from 1990, when the Thai government started liberalisation, until the 1997 crisis.

First phase, 1915–59

Siam Cement set up its first factory in the northern part of Bangkok, which was considered appropriate because Bangkok was expanding its suburban area in that direction. However, raw materials had to be brought to Bangkok by train and ship from Lop Buri province, 153 kilometres from Bangkok. Siam Cement bought Danish production machines and brought in Danish management to start production in 1915 with a capacity of 20,000–24,000 tons per year before increasing the capacity to 120,000 tons per year in 1929.

In 1939, Siam Cement decided to build a second factory in Saraburi province (108 kilometres from Bangkok), where it found raw materials. The Second World War delayed expansion, and the second factory was only completed in 1946. Total domestic cement production was increased to 400,000 tons per year, but was still not sufficient to meet the increased demand caused by increased public infrastructure construction projects after the war. Thailand had to import about 6 per cent of demand for local use. To ensure that Thailand would have sufficient cement supply for domestic use, the government permitted another three cement companies to be set up during 1952–8. The industry was also well protected by tariffs of up to 30 per cent, which made it virtually impossible for imported cement to compete with the local product.

Second phase, 1960–73

Thailand had its First Economic Development Plan from 1961. The cement and construction materials industry was not identified for promotion because the government considered it an old and established business. One objective of the First Economic Plan was to encourage new industries to substitute their products for imports. However, the cement industry benefited greatly from economic growth during the 1960s and 1970s. Agriculture growth was 4.9 per cent, as compared to 8.5 per cent in industry, 6.9 per cent in transportation and 9.1 per cent in construction (NESDB, n.d.). The American bases in Thailand and other facilities used during the Vietnam War contributed to the construction boom and the rapid growth of the cement and construction materials industries. During 1960–8, construction by the public and private sectors increased 34.5 per cent, and the demand for cement increased 22.6 per cent annually on average (NESDB, n.d.). In addition, population growth at 3.0 per cent for the whole country or 3.9 per cent in Bangkok increased demand for house building, while cement utilisation per person increased from 68 kilograms in 1970 to 115 kilograms in 1985.

Third phase, 1973–89

The 1973 oil crisis affected the cement industry because the prices of raw materials, especially oil for production, rose dramatically. Producers and dealers took the opportunity to control cement stocks and increased the cement price by 13.3 per cent (Pudsadee, 1978: 4–31). The government introduced an anti-profiteering measure by controlling prices from February 1974. Cement producers started exporting cement to make more profits. The government then issued export controls, leading to a dispute between producers and price control regulators, which led to a reduction in business and municipal taxes from 5.5 per cent to 1.65 per cent. Negotiations between the government and producers between February and June 1973 resulted in the price rising by 10 per cent after the producers justified the price increase, especially the considerable rise in the electricity price (Pudsadee, 1978: 4–35; Siam Cement Group, 1985: 94).

During the oil crisis, the government adopted a series of short-term cement policies such as export promotion, export control, retail price control, and so on. The lack of a long-term policy for the industry discouraged businesses from expanding production, with cement producers starting to lose profits. Siam Cement experienced its first loss in 1975 (Siam Cement Group, n.d.). Faced with an uncertain market and government policy, Siam Cement focused more on diversifying its products. In fact, the company had started the strategy in 1965 as it became aware of keener competition in the cement market. In 1965, Siam Cement decided to diversify by starting the first ready-mixed concrete industry in Thailand. In addition, it provided the first ready-mixed concrete truck services in the country. Through this strategy, the company maintained control of 85 per cent of the Thai cement market.

Fourth phase, from 1990 until the 1997 crisis

From the late 1980s up to the 1990s, the Thai government, under pressure from GATT, started liberalisation. Several big companies that had depended on the domestic market and government protection measures were forced to adjust themselves in order to survive in the market, including Siam Cement.

Cement products In the early 1990s, there was a problem of cement over-supply. This was caused by the government policy of liberalising the cement industry in 1991, resulting in an increase in the number of cement producers from 3 to 8. Every factory increased its production to maintain market share. Consequently, cement supply in 1996 went up to 49.8 million tons while domestic demand was only 39.4 million tons. The surplus became more severe when the construction industry slowed down and domestic demand for cement decreased. Siam Cement's market share had also been cut from 60 to 48 per cent of the cement market (Choompon Na Lamliang, December 1996: 36). The source of income for the company derived from the cement and construction materials section was also reduced from 50 per cent to 40 per cent of total income (*Manager Magazine*

Monthly, December 1991: 135) This forced Siam Cement to adjust its strategy by increasing production efficiency, depending more on external markets and investing in neighbouring countries.

Construction materials The industry had also been affected by trade liberalisation. Siam Cement tried to diversify its markets overseas by setting up a trading company in the US and Indonesia to distribute ceramics and gypsum. In the US, Siam Cement made a joint venture with Tile Cera, Inc. to produce and sell ceramics.

Pulp and paper Siam Cement controlled 50 per cent of the Thai paper market. In the early 1990s, the company faced an anti-pulp plantation movement led by environmental groups; therefore, the government had to ban its plantation projects and most of the raw materials for paper had to be imported. Consequently, backward integration in the industry is extremely difficult.

Auto parts This industry had been affected by the government's reduction of tariffs on imported cars in 1991. As a result, the price difference between imported and locally assembled cars was only 3 to 7 per cent. Several auto parts factories had to close down because of the decreased demand for locally assembled cars. Some in the Siam Cement Group, especially A.B. Siam Battery Company experienced losses. However, Siam Cement changed its strategy by depending more on heavy machinery for construction and less on the auto parts industry.

In the 1990s, Siam Cement adjusted its strategies to depend more on the international market. This trend can be seen in the export performance for 1996, which rose 21 per cent from the previous year ('Siam Cement Group in the Liberalisation Era', *Thurakit Kao Naa*, December 1996: 42). From 1995, Siam Cement showed interest in investing abroad. Decreased demand for cement and construction materials in the domestic market and the greater demand for these products in neighbouring countries encouraged Siam Cement to invest overseas. The amount of foreign investment by the Group is valued at 50,000 baht (US$2,000 million) in the areas of cement, ceramics, sanitary ware, plastic, pulp, gypsum board, diesel engine, ready-mixed concrete, asbestos cement roofing tiles, and other products in more than ten countries in Asia, Europe and America. However, the end of the boom and property bubble from 1995–6 and the 1997–8 crisis brought an end to this phase, with the future of the conglomerate uncertain since then.

Siam Cement's success factors

Siam Cement was named as having one of the world's top industrial production management companies by *Fortune* magazine in the 1980s. This accolade is valid when one looks at the company's performance over the last century. Siam Cement is involved in 112 companies in 69 industries. The consistently high value of Siam Cement Public Company Group shares has reflected the success of this enterprise.

Obviously the success of the group depends on several factors including its product differentiation strategy, leadership and decision-making, corporate culture, connections and business associations.

Product differentiation strategy

The investment pattern of Siam Cement has been consistent by selecting industries with few competitors. If Siam Cement is not the only investor in any industry, the company will diversify its products using cement as the major raw material. In addition, Siam Cement tends to invest in protected industries, especially those protected through import tariffs or by a government policy establishing a barrier to entry.

Siam Cement's investment differentiation can be divided into investments in industries that use cement as a raw material and investments in industries not related to cement, subdivided into four groups, namely: electronic and auto parts businesses, pulp and paper ventures, international trading companies, and other businesses. In terms of companies, Siam Cement has invested in 112 companies, including 22 construction materials-related companies, 9 cement and fire resistant materials companies, 21 electronic and auto parts companies, 24 pulp and paper companies, 5 international trading companies, and 31 other companies.

Several factors seem to explain Siam Cement's investment strategies. For construction-related industries, the objectives of investment expansion were to use cement as raw material for production or 'forward integration' investment to substitute new products for imported goods, as in the case of the Thai Tile Company, set up in 1938, twenty-five years after the establishment of Siam Cement. At the beginning, Thai Tile depended totally on the Siam Cement management, but started its own management team in 1960 once its own staff was experienced enough to take care of company business. Thai Tile produced asbestos cement roofing tiles and roofing sheets, and later, float glasses. The new products helped Siam Cement to control the local market because it was the only company producing them at that time.

Another reason for investing in the construction-related businesses was in response to competition from the new cement companies the government allowed to be set up in the 1950s. Facing more competition after price controls had been introduced required a new strategy. Siam Cement responded by expanding into downstream cement-related businesses including concrete roofing tiles. Since the new industries used cement as their major raw material, they could solve Siam Cement's over-supply problem. Selling cement to these companies gave higher profits and greater flexibility to Siam Cement than cement exports.

It is noteworthy that all the industries in which Siam Cement decided to invest received the government protection in terms of import tariffs and other investment promotion measures. For instance, cement received 30 per cent tariff protection; concrete roofing tile 40 per cent; fire resistant materials 35 per cent, and so on. Also, of these industries, Siam Cement was the only one in the market, so it could control the whole market (Boonakiet, 1993: 80, 252).

Such forward integration of Siam Cement for the construction-related businesses that used cement as raw material can be summarised as follows:

- 1938 asbestos cement tiles
- 1938 asbestos cement tubes
- 1952 pre-stressed concrete
- 1953 fire resistant materials
- 1963 ready-mixed concrete
- 1970 concrete roofing tiles.

Siam Cement also invested in construction-related businesses that did not use cement as raw material but enjoyed government protection, e.g. the steel bar industry. In Thailand, shortages of construction materials – such as steel bars – had been overcome by importing. The shortages became severe during the Second World War, when transportation was extremely difficult and risky. In 1966, Siam Cement was the first Thai company to invest in steel bar production. In 1967, it set up Thai Navaloha Company to produce iron of different types such as steel bars, wire rods, etc. to supply the mining, sugar, cement, auto parts, and engine industries. In 1969, the company installed an automatic machine, the first of its kind in the country, to produce diesel engines and other auto parts to fulfil the orders (Siam Cement Group, 1985: 35). The iron casting industry had been protected by the government through a 25 per cent import tariff and 10 per cent duty reduction (from 25 per cent) for the imported raw materials for production.

In addition, Siam Cement invested in other businesses, such as plastic materials for construction, for example, PVC tubes protected by the government through a 60 per cent import tariff. As the only producer in 1970, Siam Cement enjoyed the monopoly privilege until 1991, when another five PVC companies were formed. Siam Cement's investment in construction-related industries that do not use cement as a raw material can be summarised as follows:

- 1966 construction steel bars
- 1968 fibre glass roofing sheets
- 1970 PVC tubes
- 1976 cement bags
- 1979 floor/wall tiles
- 1982 gypsum boards
- 1984 plywood doors
- 1985 sanitary ware
- 1987 fitting machine sales and services
- 1991 glass sheets
- 1991 glass fibre tubes.

In sum, the pattern of Siam Cement's investments during the past eighty-five years involves forward and backward integration and diversification into industries highly protected by the government; this contributed to the growth and success of the company. However, this was not the only factor. The decision-making system, leadership, corporate culture, connections, research, human resource development, and long-term strategy have also been factors contributing to the company's international competitiveness.

Leadership and decision-making

In a large organisation like Siam Cement, leadership has been a very important factor for company success. However, it is almost impossible to determine who controls the decision-making mechanism and direction of Siam Cement. Since 1978, when Siam Cement became a public company, decisions have been made by or through the company's management committee consisting of seven managers – four from Siam Cement (general manager, vice-general manager and two assistant general managers), the senior assistant general manager of the construction group; the senior assistant general manager of the machine and electronics group; and the senior assistant general manager of the pulp, paper and petrochemicals group. Any decision, such as a 20-million baht investment, has to be screened and approved by this committee (interview with a Siam Cement official, 20 February 1997).

The committee system started in 1974 when Siam Cement had its first Thai general manager. For sixty-one years (1913–74), the company had had Danish general managers. Lacking skills in management and technology forced the company to depend on foreign managers who helped develop its modern accounting system, expand the company's product diversification, and access international finance for Siam Cement such as the World Bank's International Finance Corporation (IFC) that approved a loan of US$18 million without government guarantee (Siam Cement Group, 1985: 38–9).

In 1972, Siam Cement started its restructuring and decision-making system. In the past, each associated company under Siam Cement had its own decision-making through the company committee. Such a system caused some problems, especially in terms of co-ordination and redundancy. Siam Cement corrected the mistakes by concentrating the administration and control system at the centre under the Siam Cement board, which made long-term plans for the Siam Cement Group. Each associated company could operate its own production but was expected to report to the board. Through this mechanism, Siam Cement centralised group decision-making and decentralised operations to the associated company level.

The first two Thai general managers from 1974 had held high-ranking positions in the government. Boonma, the first Thai general manager, had been Permanent Secretary of the Finance Ministry, and his successor, Sommai was a former Assistant Governor of the Bank of Thailand. Sommai left Siam Cement to take a political position as Finance Minister in the Prem Government (1976–80). Sommai had recognised the negative impact of the oil crisis for the cement industry. Once in charge, he pursued industrial diversification so that Siam Cement would not have to rely too much on the construction industry. During his leadership, Siam Cement diversified its business from construction materials to pulp and paper, machinery, trading, as well as the automotive, petrochemical and electronics industries (Suehiro, 1996: 239–44).

Consequently, Siam Cement now comprises six major groups, collectively called the Siam Cement Group (SCG). Sommai set up a new board of directors, which represented SCG as a group, and not only Siam Cement. In addition, he upgraded SCG into a higher technology producer, e.g. the diesel engine for agricultural purposes for which Siam Cement received production technology from its Japanese partner, Kubota. In the past, Siam Cement had bought technology from the West, but this was the first time the company bought from Japan, perhaps due to Sommai's educational background in Japan. Sommai was especially interested in human resource development, and under his leadership, Siam Cement officials were sent to Japan, South Korea and Taiwan to learn new management technology.

One of Sommai's followers, Charas Xuto, was promoted to his position after years of working with the company, the first time an official of Siam Cement was promoted to general manager. Charas emphasised human resource development and management by participation. Siam Cement's general managers since Charas have been promoted from within the company. Paron, his successor, continued human resource development as well as business expansion. Choompon is the youngest general manager Siam Cement has ever had, and was only 46 years old in 1993 when he took the position. Compared with Paron, who was much older and more conservative, Choompon seemed more aggressive, judging by his past performance, when he took over the Siam Craft Company (a paper company) in the mid-1980s. His fast decision-making style probably suits the recent situation when Siam Cement wanted to increase its international competitiveness (interview with Choompon, February 1993).

Corporate culture

Siam Cement is known for its product quality, human resource development and conservative management style. In terms of product quality, Siam Cement selects from the world's best technologies for its production processes. For instance, it has bought new technology for ceramic production from Japan and Italy. This gave Siam Cement control of 29 per cent of world ceramic production. In other industries, such as iron and steel, electronics, tyres and diesel engines, Siam Cement has also depended on its partners to provide new production technology. As a result, many Siam Cement products have received ISO 9002 certification to guarantee product quality.

With regards to human resource development, Siam Cement has invested in training its staff from the early phases of the company's development. Human resource development has included in-house training, in which teams from famous institutions such as Harvard University are invited to conduct special training programmes in Thailand. Siam Cement also provides a number of scholarships for its middle- and high-ranking officials to enter degree or non-degree programmes arranged by national universities.

Siam Cement is known for its conservative management style, which can be seen from the way the company makes decisions on its investments. The SCG committee agreed that it would not diversify its investments beyond the areas already covered: cement, fire resistant materials, construction materials, machines and engines, pulp and paper, petrochemicals. This self-imposed constraint comes from the Siam Cement tradition of not 'buying' personnel from other organisations. Instead, Siam Cement tends to train its own staff. The company has developed life-long employment scheme with excellent welfare benefits. The Siam Cement workers are loyal to the company and the turnover rate is very low. At the same time, in this stable and secure environment, Siam Cement workers seem to be conservative and less dynamic, which makes it more difficult to adjust to new more competitive environments.

Connections and business associations

The leadership of the Siam Cement Group has played a significant role in Thai industrial policy. Several points should be noted in this regard:

- In the past, Siam Cement was respected by the government and the people because the Crown Property held most of its shares, and it was looked upon as a 'royal' company. The first generation of the company's general managers were Danish who helped develop professional administration. The import-substitution policy benefited Siam Cement through taxes, tariffs and entry restrictions. These policy measures were straightforward, and the company did not need to be involved directly in the policy process.
- With the second generation of management, when Siam Cement's general managers were selected from high-ranking bureaucrats (Boonma and Sommai), one sees close connections between Siam Cement leaders and industrial policy-makers. Boonma, as former Permanent Secretary of the Finance Ministry, and Sommai, as former Assistant Governor of the Bank of Thailand, could access and 'talk' with policy-makers on tax, tariff and protection measures affecting them. For instance, when the government set price controls for cement in 1974, Boonma talked with the Ministry of Commerce and influenced the government's price control rates (Siam Cement Group, 1985: 71–3). Also, Siam Cement had a group of advisers consisting of former high bureaucrats such as Vicha Setabutra, former Director General of the Mining Department, Jang Ratanarat, former Permanent Secretary of Industry, and so on. The linkages between Siam Cement and the government were obviously close, helping the company to access the policy process and protect its interests easily.
- The third or present generation of management consists of general managers who have been recruited from among the company staff. These people have long been involved in Thai business associations such as the Federation of

> Thai Industries (FTI), the most important business association. Siam Cement's key personnel hold FTI positions and are very active in their dialogues with the government. Paron, general manager of Siam Cement was honorary FTI chairman in 1987–90. Siam Cement personnel also hold positions in clubs under the FTI umbrella, in which the SCG has business interests. For instance, Alongkot Chutinan of the Siam Nawaloha Foundry Co. Ltd was President of the Auto Parts Industry Club in 1995–6; Kanok Pongpipat of the Siam Iron company was President of the Thailand National Iron and Steel Industry Club in 1995–6; Prasit Chansithichok of Thai Petrochemical Industry Public Co. Ltd was Vice Chairman of the Plastic Industry Club; Adul Udol of Siam Kraft Industry Co. Ltd was President of the Pulp and Paper Club; Thavee Butsunthorn of Siam Cement was President of the Thailand Fellowship of the Cement Manufacturer Club (Federation of Thai Industries, 1996).

It is obvious that linkages between Siam Cement managers and government policy-makers gave them access to information and people in power in order to protect and advance the interests of their companies. How did such institutional factors affect SCG performance? Our study shows that SCG benefited from institutional features at government, business-interest association and firm levels.

At the government level, the Thai state has provided protection for industries in which the SCG is involved, for example through taxes, tariffs and market access for new investors, thus contributing to SCG's growth. Government protection involved conditions for SCG to expand its product range and move from simple to higher value-added production. Although SCG received a high level of protection, it did not neglect human resource investments which helped upgrade the group. SCG leaders' personal networks and ties, especially those of Sommai and his Japanese connections, assisted the group to obtain resources, such as capital and technology, not available in the free market. Chulalongkorn University alumni networks also helped SCG to recruit staff. Professional management, rather than family rank, as well as its corporate culture strengthened the quality of SCG teamwork.

Unlike the case of the gems and jewellery companies that make use of the business-interest associations to solve their collective problems in areas such as training personnel and sharing information on international markets, SCG is less dependent on such links since it is self-reliant in training its own personnel and human resources. SCG does not obtain technical or other types of support from the associations because it has its own technological resources, derived mostly from its Japanese and foreign joint ventures. For SCG, the associations represent the industries in seeking to influence the public policy-making process and policy implementation.

The first function is more important for SCG, that is, to influence state policy, e.g. taxes, tariffs and quotas. Since SCG personnel hold key positions in the FTI clubs, they can lead the clubs in ways to benefit SCG. For instance, in 1973, when there was a shortage of cement products in the domestic market, the government adopted a price control policy and allowed new investors to enter the cement industry. Siam Cement could 'talk' with the government then to make compromises in its interest. In the early 1990s, as a result of pressure from GATT for Thailand to liberalise its steel and petrochemical industries, the government was forced to liberalise these markets. Big business groups such as SCG made demands on the government directly or through the business association channels to protect their interests. The FTI thus slowed down the pace of liberalisation of the steel and petrochemical markets.

At the firm level, institutional factors have affected SCG performance. Siam Cement also internalised transactions within its own group, allocating capital to diversify its activities. It thus linked primary inputs and intermediate goods produced by group firms such as cement and concrete roofing. In addition, participation in multiple activities has allowed SCG to reduce uncertainty concerning investment and production decisions.

Conclusions

How do institutional factors explain the growth of the two industries? This study shows that such growth can be analysed in terms of official and unofficial institutional factors. At the official level, we have seen flexible relations between businesspeople and state officials. Such relations have solved collective action problems in areas such as inadequate skilled personnel, export promotion and information for the gems and jewellery industry. Other areas that have become crucial for increasing international competitiveness like establishing brand names, R&D and securing raw materials have been brought onto the agendas of the government committees whose members include representatives of government agencies and business associations.

For the SCG, official institutional factors, especially government protection, policies and measures have contributed to the growth and success of the group. At the same time, SCG made use of unofficial institutional factors such as personal networks to obtain investment capital and upgrade technology through foreign joint-venture arrangements. Corporate networks involving university alumni links have assisted SCG in recruiting engineering staff. Such unofficial factors have helped SCG solve problems of capital, labour shortage and technology not available in free markets.

The gems and jewellery industrial group and Siam Cement Group are not first-class international exporters. Their weaknesses can be seen in their limited capacity to develop their own technology, which is essential if they want to compete at the international level. This weakness has been partly caused by government protection policies in the past in the case of SCG and lack of rational export promotion policies in the case of the gems and jewellery industry. International competitiveness in future will depend on how the Thai state and businesses effectively develop institutional responses to solve the weaknesses in the industries.

Finally, the effects of the economic meltdown in Thailand on the gems and jewellery industry and the Siam Cement Group in the late 1990s should be mentioned. As a result of chronic current account deficits, the virtually fixed exchange rate regime of the Thai currency to the US dollar, the absolute fall in the export growth rate, the failure of several finance companies in Thailand due to non-performing loans, the rise of the US dollar from mid-1995, and the massive speculative attacks on the Thai baht, the Thai authorities' attempt to defend its fixed exchange rate became increasingly untenable and impossible.

The adverse effects of the economic collapse have already been felt throughout the economy. The drastic devaluation of the Thai currency has caused price increases, especially of items having high import content, fuelling domestic inflationary pressure. Thailand's comparative advantage should improve due to the fall in the baht's value (at least against China and India which have taken over Thailand's dominance of low-priced garment and textile products), but until late 1997, the export performance had not improved much, perhaps because the exchange rate situation was still very unstable, due to political uncertainty and lack of confidence in the government.

What has happened to the gems and jewellery industry and the Siam Cement Group? With the baht falling so much in value, revenue in dollar terms will decline in the short-run despite the likelihood of increased export sales. Nevertheless, the gems and jewellery industry should recover quickly, especially with foreign tourism. The industry also received an added bonus when the government exempted it from the VAT increase from 7 to 10 per cent in early August 1997. The only major problem is that if the industry is to move up to high-end gems and jewellery products, the high costs of acquiring the machinery and technical know-how may discourage this in the immediate future.

As for the Siam Cement Group, the drastic depreciation of the baht caught the company off guard. It had incurred large dollar loans without hedging adequately for exchange rate risks. Therefore, in the space of a few days, it saw its foreign debt rise by more than 50 per cent. The many years of living under a virtual fixed exchange rate regime have cost the Siam Cement Group, and many other otherwise well-run companies in Thailand, dearly in terms of foreign exchange losses.

Epilogue: economic crisis

With the 1997–8 economic crisis, both the Siam Cement Group as well as the gems and jewellery industry had to adopt several strategies to survive in the business world. How have institutional factors contributed to the sustainability of these two groups during the toughest period of recent Thai economic history?

The gems and jewellery industry

The economic crisis in Thailand became very serious in 1997 with economic contraction and drastic devaluation of the baht. However, the economic impact of the crisis on the gems and jewellery industry was relatively minor because the industry is export-oriented, with about 80 per cent of production exported and the rest serving domestic demand. Besides, and more importantly, the foreign debt burden of the industry was quite low at the beginning of the crisis.

Nevertheless, during 1997–8, when the crisis was at its worst, export values in terms of US dollars actually declined, though it increased slightly in baht terms due to the drastic devaluation (see Table 6.1). The main factor contributing to this was the economic slowdown in the main importing countries in the region. The performance of the industry recovered strongly in the year 2000 in terms of both export values as well as domestic demand. Economic recovery and close co-operation among government, business institutions and the private sector have been the main contributors to the strong recovery of the industry. During the worst period of the economic crisis, government policies and measures also sought to increase the competitiveness of the industry. Co-operation between the government and the private sector led to measures that helped promote exports and increase domestic demand. The main measures were:

- Restructuring of the tax system for gems and jewellery. In order to increase the industry's competitive edge, the government lifted import taxes on 11 raw material items, e.g. pearls, silver, platinum. Exports of gems and precious stones were exempted from VAT, with VAT paid reimbursed to tourists leaving the country.
- To facilitate exports, the Ministry of Commerce set up a one-stop service centre to expedite export procedures. In addition, the Ministry of Commerce, the Thai Chamber of Commerce, and the Gems and Jewellery Committee sent a high-level trade mission to South Africa – the main source of gems and precious stones in Thailand – to strengthen long-term trade relations.
- The Cabinet also agreed to set up the National Institute of Gems and Jewellery Research and Development in September 1998, and the Institute started service in 1999. The main services of the Institute include: gems inspection, issuance of quality accreditation to producers, setting up a data bank and network for gems and jewellery, and creating computer software for jewellery design. The setting up of the data bank and network, especially for e-commerce, is expected to extend and expand the market for the industry.

The Siam Cement Group[5]

The 1997–8 economic crisis had a more serious impact on the Siam Cement Group than on the gems and jewellery industry. In contrast to the gems and jewellery industry, which depends heavily on the export market and less on foreign debt, the SCG faced a dilemma since the weak baht helped exports, while its accumulated offshore loans also ballooned to US$4.2 billion. Both its exports and overseas borrowings were large. Only one week after the baht floated in July 1997, the SCG lost 5.51 billion baht (at an exchange rate of 28.62 baht to the US dollar at that time). It was estimated that if the baht had gone then to 34–35 baht to the US dollar, the losses would have risen to as high as 12.5 billion baht.[6]

Table 6.2 Siam Cement Group performance, 1997–2000

	1997	*1998*	*1999*	*2000*
Revenue (bn baht)	122.6	113.2	101.9	128.2
Net profits (bn baht)	–0.05	0.02	–4.8	0.04
Foreign debt (US$)	4.2bn			840m

Source: Various Siam Cement Group Annual Reports

After July 1997, the SCG developed several strategies to maintain its empire. Four years later, in early 2001, analysts said the group's performance figures (Table 6.2) showed that the SCG had clearly passed its lowest point. The group had reduced its debt burden and refinanced its overseas loans.[7] Why and how could the SCG survive in spite of the poor performance of other big conglomerates? The SCG strategies can be described as follows:[8]

- *On foreign debt*
    - talking with Japanese, European and American lenders to roll over repayment of US$1 billion in short-term loans;
    - withholding dividend payment in 1997 for the first time in more than two decades;
    - rescheduling repayment, but the SCG did not renege on loan repayments because it would give the group a negative image in the eyes of lenders and lose its good credit standing for the future;
    - converting about 30 per cent of its foreign debt into yen.
- *Cutting expenses*
    - deferring all spending on expansion and new projects;
    - stopping dividend payments to shareholders;
    - stopping bonus payments to the group's directors while staff bonuses ranged between four to six weeks rather than the previous two to three months.
- *Raising revenues*
    - increasing cement prices to cover production costs;
    - selling a large stake in non-core businesses to its partners such as the auto parts affiliates, Thai Engineering Products Ltd, Navaloha Industry, Siam AT Industry, Thai CRT Co., Siam NEC Co., Siam Hitachi Machinery Co., Siam Magotteaux Co., Siam Fuchs Lubricants, etc.;
    - raising exports;
    - issuing 400 million baht of shares;
    - allowing foreigners to own 40 per cent of its shares, up from 25 per cent in the past (this strategy was part of a planned 400-million baht share issue);

  - hiring independent valuers to assess its assets, especially imported machinery, to reflect current market prices; consequently, the exercise added value totalling 90.693 billion baht to the group's financial position.

- *Investment*
  - opening another Siam Cement trading office, its twelfth, in Dubai, part of the United Arab Emirates, serving trade with Europe and Africa;
  - dropping plans to seek foreign partners who would provide technology and marketing expertise because it had revised its marketing strategy and would now focus only on domestic sales, as well as sales to Indonesia and South Asia.

- *Restructuring*
  - dividing its major businesses into six groups and two holding companies for investment and property management purposes. Of the six business groups, the three main units were cement, paper and packaging, and petrochemicals; three smaller units handle building products, ceramics and trading. The purpose of the restructuring was to achieve a clearer corporate structure, enhance efficiency and resource management, while strengthening operations through long-term strategies. This model replaced the 1999 business structure, which divided operations into core, potential and non-core businesses.

- *Technology*
  - upgrading its cement-making machinery to increase exports from 4.8 million tons in 1999 to 7 million tons in 2000, with factories running at 66 per cent of total capacity in 2000.

Notes

1 Anek pointed out that the relationship between the state and organised business in Thailand was not unproblematic. A major criticism of this relationship was that it kept the popular sectors out of the economic policy-making process. Despite intermittent calls for the inclusion of farmer and worker groups in the Joint Public–Private Consultative Committee (JPPCC) – an influential body where three of the most important business associations in Thailand, namely the Federation of Thai Industries (FTI), Thailand's Board of Trade (BOT), and the Thai Bankers Association (TBA), could exert influence on the government's policy-making – senior officials of the National Economic and Social Development Board (NESDB), the secretariat of the JPPCC, insisted that such inclusion was undesirable. Another relevant criticism was that government–business collaboration had focused almost exclusively on the reduction of government-caused costs so that Thai businesses could compete more effectively. Environmental degradation or how to correct the highly skewed income distribution was simply left out of consideration in government–business dialogues.

2 While Sakkarin is not an uncritical believer in the bureaucratic polity formulation, he emphasises the important role of the modern bureaucracy in current economic policy-making. Since the bureaucracy is still the most highly organised public institution,

controlling information and most areas of society, it has remained influential despite its reduced role in the political system. The actual interplay among private business, the bureaucracy and political power in an increasingly globalised world have led to what Sakkarin calls a 'liberalisation coalition'. He uses the case of telecommunications liberalisation in Thailand from the early 1980s to the early 1990s to support his argument about a 'liberalisation coalition' in economic policy-making.

3 See Medhi Krongkaew, 'The economics of institutional change: the private sector and the making of economic policy in Thailand' in Colin Barlow (ed.) *Institutions and Change in Southeast Asia*, Edward Elgar, London, 1999.

4 While it is argued that economic prosperity, which involved the rise of the *nouveau riche*, erodes the power and prestige of poorer civil servants, the current economic situation, where the private sector is suffering from the economic slowdown, may see the resurgence in the power and prestige of civil servants or public bureaucrats.

5 The authors would like to express their appreciation to Mr Bukbhun Boonlert of the *Matichon Daily* (Thailand) for data on the Siam Cement Group after the crisis.

6 *The Bangkok Post*, 28 August 1997.

7 Interview with Sukit Udomsirikul of Capital Nomura Securities, quoted from the *Bangkok Post*, 27 April 2000.

8 Data and information in this section and the section on performance are derived from several newspapers, including *Matichon Daily*, *Prachachat Thurakij*, *Krungthep Thurakij*, *Pujatkarn Daily*, *The Bangkok Post*, and *The Nation*, during the period July 1997–January 2001.

References

Anek Laothamatas (1989) 'Business and politics in Thailand: new patterns of influence', *Asian Survey*, 28: 4.

—— (1992a) 'Business on the democracy road: theory and reality', paper presented at the Third Annual Conference on the 'Middle Class and the Development of Democracy in Thailand', organised by the Political Economy Study Centre, Faculty of Economics, Chulalongkorn University, Bangkok, 18–19 November (in Thai).

—— (1992b) *Business Associations and the New Political Economy of Thailand: From Bureaucratic Polity to Liberal Corporatism*, Singapore: Institute of Southeast Asian Studies.

Bangkok Gems and Jewellery, October 1995.

Bank of Thailand (BOT) (1992) *Industrial Business Situation in 1991* (in Thai).

—— (1993) *Industrial Business Situation in 1992 and Trends in 1993* (in Thai).

Boonakiet Ketkaomaneerat (1993) 'Investment differentiation of Siam Cement Group,' MA thesis, Faculty of Economics, Thammasat University, Bangkok.

Chai-Anan Samudavanija (1987) 'The bureaucracy' in Somsak Xuto (ed.) *Government and Politics of Thailand*, Singapore: Oxford University Press.

—— (1988) 'Overview of the determination of Thai economic policies', in *Who Determines Thailand's Economic Policies?*, Bangkok: Faculty of Economics, Thammasat University, pp. 64–107 (in Thai).

Choompon Na Lamliang (1993) 'Siam Cement's new direction', interview in *Thurakit Kao Naa* (Business in Progress), February.

—— (1996) 'Today', interview in *Thurakit Kao Naa* (Business in Progress), December.

Christensen, Scott R., David Dollar, Ammar Siamwalla and Pakorn Vichyanond (1992) 'Institutional and political bases of growth-inducing policies in Thailand', draft paper prepared for the World Bank project on the 'Role of Government and East Asian Success', October.

Department of Export Promotion (DEP) (1996) *Gems and Jewellery Export Situation January–June* 1996 (in Thai), Trade Document Section, Trade Advisory Division.

Doner, Richard (1991) 'Approaches to the politics of economic growth in Southeast Asia', *Journal of Asian Studies*, 50(4): 818–49.

—— (1992) 'Politics and the growth of local capital in Southeast Asia: auto industries in the Philippines and Thailand', in Ruth McVey (ed.) *Southeast Asian Capitalists*, Ithaca, New York: Cornell Southeast Asia Program, pp. 191–218.

Doner, Richard and Ansil Ramsay (1989) 'Thailand as a case of flexible strength', paper prepared for the Annual Meeting of the Association for Asian Studies, Washington, DC, 17–19 March.

Doner, Richard and Patcharee Siroros (1993) 'Contending with fragmented markets: firm and group in the Thai auto parts industry', unpublished paper.

Doner, Richard and Patcharee Siroros (1995) 'From local content to liberalisation: the state and business in the Thai automobile industry', paper presented at the Georgetown Conference on Korean Society II, 'Korean State and the Rise of Civil Society: Comparative Perspectives', organised by Georgetown University, Washington DC, 7–10 May.

Federation of Thai Industries (FTI) (1996) *Federation of Thai Industries Directory 1995/1996*, Bangkok: Federation of Thai Industries.

Frey, Bruno S. (1983) *Democratic Economic Policy: A Theoretical Introduction*, New York: St Martin's Press.

Girling, John (1981) *The Bureaucratic Polity in Modernising Societies: Similarities, Differences and Prospects in the ASEAN Region*, Singapore: Institute of Southeast Asian Studies.

Grit Permtanjit (1982) *Political Economy of Dependent Capitalist Development: Study on the Limits of the State to Rationalise in Thailand*, Bangkok: Chulalongkorn University Social Research Institute.

Hawes, Gary and Hong Liu (1993) 'Explaining the dynamics of Southeast Asian political economy: state, society, and the search for economic growth', *World Politics*, 5(4), July.

Hewison, Kevin (1985) 'The state and capitalist development in Thailand', in Richard Higgott and Richard Robinson (eds) *Southeast Asia: Essays in the Political Economy of Structural Change*, London: Routledge and Kegan Paul.

—— (1989) *Bankers and Bureaucrats: Capital and the Role of the State in Thailand*, Monograph Series 34, Yale University Southeast Asia Studies, Yale Center for International and Area Studies, New Haven, CT.

Industrial Corporation of Thailand (ICT) (1995) *Industrial Business Situation in 1994 and Trends* (in Thai).

—— (1996) *Industrial Business Situation in 1995 and Trends* (in Thai).

Industrial Finance Corporation of Thailand (IFCT) (1986) *Export Industries* (in Thai), Business Research Center, Research and Planning Division.

Industrial Management Company (IMC) (1991) *Structure of Gems and Jewellery Study Project* (in Thai), Report to Department of Export Promotion, August.

Kakwani, Nanak and Medhi Krongkaew (1997) 'Some good news on poverty and income inequality', *Bangkok Post Year-end Economic Review*, 1 January.

Klein, C. and C.S. Hurlbut, Jr (1985) *Manual of Mineralogy*, twentieth edn, New York: John Wiley & Sons.

Kraisak Choonhavan (1984) 'The growth of domestic capital and Thai industrialisation', *Journal of Contemporary Asia*, 14(2): 135–46.

Krirkkiat Phipatseritham (1993) 'Privatisation policy for public enterprises', in *Fiscal*

Revolutions for Economic Reforms, Faculty of Economics, Thammasat University, Bangkok.

Krueger, Anne O. (1974) 'The political economy of the rent-seeking society', *American Economic Review*, 64 (June): 292–303.

Lin, Justin Yifu and Jeffrey B. Nugent (1992) 'Institutions and economic development', in J.R. Behrman and T.N. Srinivasan (eds) *Handbook of Development Economics*, vol. IIIA, Amsterdam: North-Holland, 1995.

MacIntyre, Andrew (1990) 'Business–government relations in industrialising East Asia: South Korea and Thailand', Australia–Asia Papers no. 53, Centre for the Study of Australia–Asia Relations, Griffith University, Nathan, December.

Mackie, J.A.C. (1992) 'Changing patterns of Chinese big business in Southeast Asia', in Ruth McVey (ed.) *Southeast Asian Capitalists*, Ithaca, NY: Cornell Southeast Asia Program, pp. 161–90.

Manager Magazine Monthly (1991) 'Yang Ni Tong Yang Naa' (Siam Cement's advertisement of its products), interview column, December.

McVey, Ruth (ed.) (1992) *Southeast Asian Capitalists*, Ithaca, NY: Cornell Southeast Asia Program.

Medhi Krongkaew (1988) 'Economic theories and the making of economic policies: applications in the Thai economy', in *Who Determines Thailand's Economic Policies?*, Bangkok: Faculty of Economics, Thammasat University, pp. 1–63 (in Thai).

Medhi Krongkaew, Pranee Tinakorn and Suphat Suphachalasai (1992) 'Rural poverty in Thailand: policy issues and responses', *Asian Development Review*, 10(1): 199–225.

Ministry of Industry (1996) *Industrial Master Plan for Thailand*, Bangkok (in Thai).

—— (1997) 'Industrial restructuring', Bangkok, September (in Thai).

Muscat, Robert J. (1990) *Thailand and the United States: Development, Security, and Foreign Aid*, New York: Columbia University Press.

National Economic and Social Development Board (NESDB) (1984) *Gems Industry in Thailand* (in Thai), Industrial Planning Section, Economic Projects Division.

—— (n.d.) GDP Annual Report. Accountancy Division.

Nilphechara, K. (1982) *Supply Study: Gems and Jewellery in Thailand*, Department of Commercial Relations, Thailand.

North, Douglas (1990) *Institutions, Institutional Change and Economic Performance*, Cambridge: Cambridge University Press.

Olson, Mancur (1965) *The Logic of Collective Action*, Cambridge, MA: Harvard University Press.

—— (1982) *The Rise and Decline of Nations: Economic Growth, Stagflation, and Social Rigidities*, New Haven: Yale University Press.

Pasuk Phongpaichit (1992) 'The role of middle class in economics and politics of the Asian NIC countries and Thailand: preliminary observations', paper presented at the Third Annual Conference on the 'Middle Class and the Development of Democracy in Thailand', organised by the Political Economy Study Centre, Faculty of Economics, Chulalongkorn University, Bangkok, 18–19 November (in Thai).

Pasuk Phongpaichit and Chris Baker (1996) *Thailand's Boom!*, Bangkok: Silkworm Books.

Patcharee Siroros (1992) 'Political regimes, Japanese investment and policy choices: a study of the Thai auto industry', *USJP Occasional Paper* 92–13, Harvard University.

—— (1997) *Thai State and Business in Automotive Industry*, Bangkok: Thammasat University Press (in Thai).

Patcharee Thanamai (Siroros) (1985) 'Patterns of industrial policy-making in Thailand:

Japanese multinationals and domestic actors in the automotive and electrical appliances industries', PhD dissertation, University of Wisconsin, Madison.
Prayong Nettayarak (1988) 'The making of agricultural economic policies: case study of the setting of sugarcane price', in *Who Determines Thailand's Economic Policies?*, Bangkok: Faculty of Economics, Thammasat University, pp. 198–331 (in Thai).
—— (1992) 'The future of Thai sugarcane and sugar industry', paper presented at the 1992 Annual Year-end Conference of Thailand Development Research Institute, Chon Buri, 12–13 December (in Thai).
Preecha Piampongsarn (1992) 'The theory of the middle class', paper presented at the Third Annual Conference on the 'Middle Class and the Development of Democracy in Thailand', organised by the Political Economy Study Centre, Faculty of Economics, Chulalongkorn University, Bangkok, 18–19 November (in Thai).
Pudsadee Hor-nak (1978) 'Thai Cement industrial development,' MA thesis, Faculty of Economics, Thammasat University, Bangkok.
Ramsay, Ansil (1987) 'The political economy of sugar in Thailand', *Pacific Affairs*, pp. 248–70.
Rangsun Thanapornpun (1989) *The Process of Economic Policy Determination in Thailand: Analysis of Political Economic History, 1932–1987*, Bangkok: Social Sciences Association (in Thai).
Read, P.G. (1991) *Gemology*, London: Butterworth-Heinemann Ltd.
Riggs, Fred W. (1966) *Thailand: The Modernisation of a Bureaucratic Polity*, Honolulu: East-West Center Press.
Sakkarin Niyomsilpa (1995) 'The political economy of telecommunications liberalisation in Thailand', unpublished PhD dissertation, Australian National University.
Schotter, Andrew (1990) *Free Market Economics*, second edn, Cambridge, MA: Basil Blackwell.
Siam Cement Group (1985) *Siam Cement, 2455–2526*, a report on the occasion of 71 years anniversary of Siam Cement Group (in Thai).
—— (n.d.) *Siam Cement Annual Report, 1972–1978.*
Somboon Siriprachai (1989) 'The survey of state-of-knowledge of rent-seeking activities and the case study of export quotas of Thai cassava to the European Community, 1981–1982', research report submitted to the Research and External Division, Thammasat University, 12 September (in Thai).
Somkiat Wantana (1992) 'Local business and Thai democracy', paper presented at the Third Annual Conference on the 'Middle Class and the Development of Democracy in Thailand', organised by the Political Economy Study Centre, Faculty of Economics, Chulalongkorn University, Bangkok, 18–19 November (in Thai).
Stigler, George J. (1975) 'The theory of economic regulation', in *The Citizen and the State: Essays on Regulation*, Chicago: University of Chicago Press.
Stiglitz, Joseph E. (1996) 'Some lessons from the East Asian miracle', *The World Bank Research Observer*, 11(2), August: 151–77.
—— (1997) 'Keynote address: the role of government in economic development', in Michael Bruno and Boris Pleskovic (eds) *Annual World Bank Conference on Development Economics 1996*, Washington, DC: The World Bank, pp. 11–26.
Stiglitz, Joseph E. *et al.* (1989) *The Economic Role of the State*, Oxford: Basil Blackwell.
Suehiro, Akira (1992) 'Capitalist development in post-war Thailand: commercial banks, industrial elite, and agribusiness groups', in Ruth McVey (ed.) *Southeast Asian Capitalists*, Ithaca, NY: Cornell Southeast Asia Program, pp. 35–64.
—— (1996) *Capital Accumulation in Thailand, 1855–1985*, Bangkok: Silkworm Books.

Sunetra Sungsuwan (1991) 'The JPPCC and the determination of Thailand's economic policies', MA thesis, Faculty of Economics, Thammasat University, in Bangkok (in Thai).

Thailand Business (1987) 'Facets of the Thai gemstone trade,' July–August.

Thiranart Karnjana-aksorn (1992) 'The military and business', paper presented at the Third Annual Conference on the 'Middle Class and the Development of Democracy in Thailand', organised by the Political Economy Study Centre, Faculty of Economics, Chulalongkorn University, Bangkok, 18–19 November (in Thai).

World Bank (1980) *Thailand: Toward a Development Strategy of Full Participation*, World Bank Country Study, East Asia and Pacific Regional Office, World Bank, Washington, DC.

—— (1997) *World Development Report 1997: The State in a Changing World*, Washington, DC: The World Bank.

Index